WELLNESS AT WORK

A REPORT ON HEALTH AND FITNESS PROGRAMS FOR EMPLOYEES OF BUSINESS AND INDUSTRY

ROBERT M. CUNNINGHAM, JR.

AN INQUIRY® BOOK 1982
Blue Cross Association
Blue Shield Association

An INQUIRY® Book

676 St. Clair Street,
Chicago, IL 60611.

First Printing: January 1982

Library of Congress Cataloging in Publication Data

Cunningham, Robert Maris, 1909-
Wellness at work.

"An Inquiry book."
Bibliography: p.
Includes index.
1. Industrial hygiene. 2. Employee assistance programs. I. Title.
HD7261.C83 658.3'82 81-15500
ISBN 0-914818-08-2 AACR2

Design by Jan Roy

Table of Contents

Foreword

Nobody ever really believed that an apple a day keeps the doctor away, as our grandmothers used to say, but as health care costs have soared in recent years, reaching a total now exceeding $200 billion a year, there is increasing evidence that Grandmother had the right idea. The combined efforts of government, providers of health care services, employers, and Blue Cross and Blue Shield Plans and other health insurers are not going to succeed in controlling health care costs unless at the same time there can be a corresponding national commitment to preventing disease and injury and promoting physical fitness and good health habits. Up to now, only an insignificant fraction—estimated to be less than 2 percent of the total amount spent for health—has been devoted to keeping people well, yet there is encouraging evidence in many studies reported here and elsewhere that investments in prevention and fitness offer returns not only in fewer and smaller doctor and hospital bills but in longer life, greater productivity and effectiveness, and improved ability to cope with the stresses of modern life.

The current administration in Washington is determined to see to it that government makes its contribution to the encouragement of the need for preventive services and healthy habits of living. And it is heartening to see in Robert Cunningham's report that providers of health care services, insurers, employers, and communities are awakening to this need and responding with new ideas and new programs. These efforts must be greatly expanded, until not just a few prudent and far-seeing corporations but every company, every employer is persuaded to take action. The work force is at the core of the nation's economy, and the leadership of employers and unions can show the way to ensuring a healthier work force, and hence a healthier nation.

But it should be emphasized again and again that the new fitness and health programs must be sustained to be effective.

Lifestyles are not changed overnight. People who have thoughtlessly but understandably allowed themselves to indulge in unhealthy habits of eating, drinking, working, and worrying can be shown how to do better and feel better. Unless healthy behavior becomes the norm, the thing to do, for the culture in which they live, however, the chances are that they will revert to their former behavior soon after the props—the nutrition film, the exercise class—are removed. So the programs and activities—some of which are sponsored by certain Blue Cross and Blue Shield Plans—aimed at changing behavior have to be sustained and repeated until the work place culture, the norm of behavior, has shifted enough to make the right things, instead of the wrong things, the things to do.

This isn't likely to happen without the wholehearted support of the nation's employers. With that kind of support, it can't fail.

Bernard R. Tresnowski
President
Blue Cross and Blue Shield Associations

Chicago, Illinois
January 1982

Chapter One
THE WELLNESS MOVEMENT

Except in the backwoods of Maine, say, or the arid wasteland of a desert in the Southwest, you can't stand on any city street or country road in the United States today for very long without seeing a determined jogger go by, feet pounding, arms swinging, face perspiring, eyes fixed determinedly on some distant goal unseen by the uninitiated. Ten years ago you could have stood in the same spot all year without seeing anything of the kind, but if you took your stand now in any population center—at a lakefront park in Chicago, or a winding, leafy street in West Los Angeles, or on the walkway around the Reservoir in Manhattan's Central Park—the joggers would pass in squads at almost any time of day.

What are they up to? They are all chasing the same thing: physical fitness. And those who have been wise enough to consult their physicians, or qualified exercise physiologists, before taking to the roads, streets, and parks, are going to catch up with their goal. Then, if they are serious enough about what they are doing to keep at it, they will soon see the absurdity of running half an hour or an hour every day and at the same time continuing to eat, drink, smoke, work, and worry themselves back into the very condition they're trying to run away from. So they will change—first their attitudes, and then their habits, and then their lives—and they will be healthier, live longer, and enjoy it more.

Nobody is going to argue with those goals, or those results, so one might think that Americans must be getting healthier by leaps and bounds, and that we'll soon be finished with our worries about the high cost of doctors and hospitals. In a nation of joggers, who's going to need them?

Of course, we aren't a nation of joggers, or anything like it. Nobody really knows how many men and women are out there jogging. The estimates go as high as 20 million, but one thing that is known is that this week's 20 million, or whatever, are not the same as last week's or next week's. Of any week's 20 million, not many will have been wise enough to consult their physicians or physiologists before they started running, so they haven't been properly instructed or equipped, and some of them may be doing themselves more harm than good. Some of these will show up at doctors' offices or clinics or hospitals complaining of shin splints, or tendinitis, or metatarsal pain, or worse. Some will just get tired of it and quit, and some who do see the absurdity of running without changing their habits of living will decide they'd rather eat, drink, and be merry than switch.

Given all this attrition, however, the numbers are still impressive, and growing, and given the certainty now that men and women who *are* adequately informed and instructed and motivated will exert extraordinary efforts to get healthy and stay that way, it should be possible to keep this "wellness movement" going and growing and make certain that it does indeed curb the rising cost of doctors and hospitals, now swollen to more than $200 billion a year. Especially, it should be possible for business and industry to turn their capability to inform and instruct and motivate their own employees toward informing and instructing and motivating them to be healthier and feel better, and cost their employers less— in dollars spent for health care and health insurance and in dollars lost to absenteeism, unproductive time, and poor work.

In fact, it's happening. Some indication of the rate at which business and industry have already turned capabilities on to those goals is apparent in the growth of the American Association of Fitness Directors in Business and Industry, an organization that started in 1975 with a membership of 25 company representatives and has 3,000 members today, notwithstanding the formidable obstacle presented by the fact that it is known familiarly as AAFDBI, or "Affdibby," by those who manage to pronounce it.

Other evidence of growth is found in the testimony of C. Carson Conrad, executive director of the President's Council on Physical Fitness and Sports, which was initiated by President Eisenhower 25 years ago and has been expanding its activities for industry in recent years. "Physical fitness programs for the work force at the work place have become a worldwide movement among developed industrial nations," Carson said in a report published in 1980. "It appears at this time that Japan is well out in front, followed by Switzerland, with the United States closing fast in third position."[1] Japan not only has the most advanced employee fitness programs but is also first in life expectancy and productivity, Conrad pointed out. "It would appear that Japanese employers have given a high priority to the role of exercise in improving the health and morale of their employees as well as increasing productivity and reducing absenteeism due to illness," he said, citing as an example an extensive fitness and sports complex of the Osaka Gas Company, whose 11,000 employees are the beneficiaries of a comprehensive program of testing, medically supervised exercise, education, and sports at the Osaka Health Services Center that includes, among other things, a gymnasium, sports medicine clinic, indoor track, swimming pool, martial arts and dance studio, weight room, and various examination, testing, education, and sports facilities. The center would remind an American of a well-planned community college physical education and athletic complex, Conrad said.

Where It Came From

To get in stride with the serious joggers and in step with all the others who are ready to join the wellness movement but don't know where or how to sign up, managers in business and industry need to know where the movement came from and what makes it go.

The wellness business— and it should be emphasized at the outset that wellness here means not simply the absence of symptoms but positive well-being, the kind of robustness that comprehends respect for the body and appreciation of its worth—didn't begin with the joggers in Central Park, or with Dr. Paul Dudley White and his celebrated bicycle at the Massachusetts General Hospital and Harvard Medical School 25 years ago, or even with Teddy Roosevelt, whose lifelong advocacy of strenuous physical exercise had our grandfathers and grandmothers patriotically inhaling deeply and doing calisthenics in front of open windows to avoid bringing up a generation of weaklings. In fact, when Asclepius built his temples of healing for the ancient Greeks of Athens, Epidaurus, Cos and Cnidus, baths, diet, and exercises were as much a part of the treatment as medicines were. And many of the temples had stadiums to encourage games and link athletics and health, a practice that is repeated in our modern health resorts with their swimming pools, tennis courts, and golf courses. Centuries later, the passion for fitness and sports was in evidence at Britain's public schools and immortalized in the Duke of Wellington's observation that the Battle of Waterloo was won on the playing fields of Eton. For obvious reasons, we prefer that example to the later demonstrations of the same principle in the Third Reich.

Over the years, then, wellness movements have emerged and subsided repeatedly, in this country and elsewhere, in response to various kinds of stimuli, having to do most often, perhaps, with what was seen as a national need or national interest or pride in strength and power. If there is a difference between the wellness movement of our time and those that have come and gone before, it is in the precise nature of the stimulus. The national need in this case has been not so much to build strength and power, desirable as these ends may be, as it has been simply to stop wasting our resources. There must be better ways to spend $200 billion, or part of it, especially in view of the fact that it is now generally agreed among health economists and public health authorities that the billions we have been adding to health care spending, year after year, are not producing corresponding improvements in the health status of the population.

Some critics of American medicine have stated the proposition more emphatically. Thus Ivan Illich, a Viennese theologian who left the priesthood to become an educator and headed a social problems think tank in Mexico, considers that medicine in

America has been "industrialized" to a point where it threatens health instead of restoring and protecting it. "Schools produce education, motor vehicles produce locomotion, and medicine produces clinical care," Illich wrote in a book called *Medical Nemesis* that elevated blood pressures in hospital corridors around the country in the early 1970s. "These outputs are staples that have all the characteristics of commodities. The main concern of medicine is sick individuals rather than the health of populations. Only a substantial reduction in total medical outputs could foster autonomy in health and in sick care, and thereby make it effective."[2]

It is possible to agree with Illich that individual autonomy in health would mean fewer medical outputs, without concluding as he did that the best way to go about achieving autonomy is to cut doctors off at the knees. "We need public and private support to test new approaches to health and new measures aimed at giving people the capability and will to take greater responsibility for their own health," Walter J. McNerney, president of the Blue Cross and Blue Shield Associations, said at a conference titled Future Directions in Health Care. "We know that health status is a compound of heredity, lifestyle, social and physical environment, and all kinds of emotional stimuli, as well as medical care, yet we continue to feed our money and manpower resources into one component, to the practical exclusion of all the others, and when the imbalance obtrudes, we have generally turned our backs and considered that redressing it is somebody else's job."[3]

Inevitably, what is somebody else's job is nobody's job until it becomes obvious that the job has to be done, and then it tends to become everybody's. That is what has happened in the case of the nonmedical components of health. Genetic scientists investigating the possibilities of recombinant DNA are getting ready to do something about heredity, not for us but perhaps for our progeny's progeny. Psychologists and sociologists and their multiple subspecialists examining our social environments are forever discovering new hazards to our emotional health in the way we look at and talk to one another, and new prescriptions to mend our behavior. The number of menaces in the physical environment is exceeded only by the number of experts leaping forward to protect us. And when it comes to lifestyle, we are all experts.

Actually, lifestyle is one of those abstractions we use all the time, usually without knowing exactly what it is that we have in mind. To some, unquestionably, it denotes the status of being married or single, at least in the legal sense, and to others it may suggest whether Sundays are spent in church or in bars, or nights out are spent at the opera or the bowling alley. In the context here, lifestyle must refer to behavior specifically related to health. While it could be argued that all behavior is specifically related to health in one way or another, the context suggests concern especially for such directly relevant components of lifestyle as diet, exercise, smoking and drinking habits. Also, importantly, "lifestyle" includes attitudes toward family responsibilities, work, recreation, and the world in general. When dedicated runners exclaim, as some do, that running has changed their whole lives, this is not to be put down as a relatively harmless form of intoxication, roughly equivalent to the second martini. It is, rather, a significant observation, and if we could somehow get more people out of bars and bleachers and onto running tracks, it might do more for the national esprit de corps than it would do for the public health.

The Breslow Prescription

Most authorities in preventive medicine agree that the most effective lifestyle changes are not all that dramatic. One of the most widely quoted and recommended regimens for healthful living was promulgated a decade ago by Dr. Lester Breslow, dean of the School of Public Health of the University of California at Los Angeles: 1) Three meals a day, no snacks; 2) Breakfast every day; 3) Moderate exercise two or three times a week; 4) Seven or eight hours sleep a night; 5) No smoking; 6) Moderate weight; and 7) Alcohol in moderation only. In a study they reported in the journal *Preventive Medicine*, Dr. Breslow and an associate kept records of the lives of 7,000 adults for a period of several years and concluded that a 45-year-old man who observed from none to three of the prescribed rules had a life expectancy of 21.6 years—to age 67, while a 45-year-old man who kept 6 or 7 of the rules would live 33.1 years, to age 78.[4]

Studies like Dr. Breslow's were the basis for an innovative

form of medical practice introduced to their patients a few years ago by Drs. Lewis Robbins and Jack Hall of the Methodist Hospital of Indiana at Indianapolis. This system, called the Health Hazard Appraisal, makes use of current morbidity and mortality tables to structure what is in effect a contract between the physician and each of his patients, in which the physician agrees to take care of the patient, and the patient agrees to abide by specific recommendations regarding diet, exercise, smoking, drinking, driving, and other pertinent habits of daily living. Referring to the tables, the physician can demonstrate how the agreed-on changes in lifestyle may lengthen the patient's expected life span by minimizing risks of heart disease, cancer, stroke, and other diseases. The contract thus provides an element—compelling motivation—that is lacking in the customary generalized admonition to "lose a little weight" or "go easy on the booze." Instead, the individual patient sees his own Health Hazard Appraisal relating the averages in the tables to his own life and his own behavior. The method has been publicized in medical and hospital journals, and several hundred physicians are now using it in their practices.

Another new approach to health care emphasizing lifestyle change is the Wholistic Health Center concept, which had its roots in early experimental trials in Ohio and Illinois and is now spreading to other parts of the country. The Wholistic Health Centers are church-based primary care facilities staffed by clergy-counselors and trained volunteers as well as family practice physicians and nurses. Except in emergencies, when referral to hospital care and specialist services takes precedence, the emphasis is on combining diagnosis and treatment with investigation of the patient's life circumstances, counseling aimed at resolving whatever problems or situations may be seen as causes of stress related to illness or symptoms, and lifestyle recommendations designed to prevent recurrence. The big idea is to consider and treat the whole person, not just the part that hurts.

The wholistic technique involves diagnostic, planning, and evaluation conferences in which the physician, nurse, counselor, and patient all take part as partners and arrive together at agreement on what is wrong and what needs to be done. With a dozen of these centers now operating, some in conjunction with medical schools, the concept is already achieving exceptional scores in patient satisfaction and compliance with recommended procedures. The founder of the centers and creator of the concept is a clergyman with years of experience as a hospital chaplain, who sees the nation's churches and trained clergy as a vast underused resource that can combine primary medical care with equal parts of emotional, social, and spiritual support, fulfilling needs that are important to good health. Granger E. Westberg, D.D., established the first wholistic clinic at the Good Shepherd Lutheran Church in Springfield, Ohio, in 1970. At the time, he had envisioned the rapid development of a nationwide network, with hundreds or thousands of these centers in communities from coast to coast. But this hasn't come to pass; acceptance of the concept has been cautious — probably for several reasons. Some people are satisfied with their dependency on physicians and are uneasy in the partnership role the wholistic concept calls for; many physicians are not ready to accept the participation of nurses, clergy, and patients in decisions they have been trained and accustomed to make by themselves, and while the healing mission is traditional in the Christian, Jewish, and other religions, the idea of seeking care in a church setting is not always acceptable.

The Idea Is Catching On

While wholism isn't sweeping the country, there can't be any questioning the fact that the big idea is catching on fast: The principle that underlies all the new ideas and programs having to do with health, from jogging to a West Coast confection called Tibetan Gong Meditation, is simply that it is a person's disposition and behavior that determine the condition of his cardiovascular and gastrointestinal systems, not the other way around. This is not a new idea. When he was chief of the internal medicine section at the Mayo Clinic 50 years ago, the late Dr. Walter Alvarez startled his colleagues and risked excommunication by remarking that he would cure most of his patients sooner if he could start by examining their bedrooms and bank accounts instead of their gall bladders and colons. The suggestion remains impractical, although most doctors today would probably acknowledge the basic truth in Dr. Alvarez's synecdoche. And in the wellness movement we are beginning, at least, to redress the imbalance that has the nation spending $200 billion for health *care* and an estimated $1 billion, if that, for health.

Wellness activity, or health promotion, as it is more commonly labeled, is in evidence all over the place. Most notably, the new ideas are stirring in the health professions, where the focus on sickness has been most entrenched in the past. Courses in preventive medicine are multiplying in medical school curricula; these are usually electives, but the astonishing thing is that growing numbers of medical students are actually electing them. Most hospitals have had educational programs of some kind for patients with heart disease, diabetes, colostomies, and other conditions requiring adjustments in daily living habits, but recently progressive institutions have been adding health education programs on a variety of topics presented on closed circuit television for viewing in patients' rooms and waiting areas. Some now have good-heath publications that are distributed not only to patients and visitors but widely throughout their communities; a few offer classes and public meetings with lectures on nutrition, exercise, stress management, and common problems like obesity, smoking, alcoholism, high blood pressure, and others. An enterprising publisher in Chicago started a health education newsletter that could be imprinted with a hospital's name and purchased in bulk for distribution in the hospital's community; in three years the circulation had grown to more than two million copies—a significant clue to the accelerated interest in wellness.

Of course, when the wellness fires started to burn a few years ago, government got into the act and set some fires of its own. City and county public health departments stepped up their health education activities for schools and community groups; lively displays and animated cartoon movies replaced the deadly pamphlet that had been standard health education fare for years. Many of the states now have active governors' councils on physical fitness and sports, organized under a White House initiative. A President's Committee on Health Education resulted in the new National Center for Health Education, supported by private funds. The center sponsors and evaluates experimental educational programs and serves as an information exchange, so A won't duplicate

B's mistakes and X can benefit from Y's success. A new Bureau of Health Education and an Office of Health Information, Health Promotion, Physical Fitness and Sports in the federal Department of Health and Human Services perform similar functions for the government.

It seems likely, however, that the cost of all the health promotion activities of doctors, hospitals, governments, and health agencies like the American Heart Association, American Cancer Society, and countless others wouldn't add up to much more than the amount already being spent by business and industry on preventive and health education and fitness programs, most of which have been initiated within the past five years. And the aggregate of existing programs can provide only a hint of industry's potential for improving the health and fitness of the work force, not to mention the impact such programs are known to have on the health of workers' families. Organized employee health or fitness programs that go beyond the usual occupational health and safety concerns of the average corporate medical department have been reported by some 400 corporations—out of thousands whose managements are worried, and in some cases panicked, about the soaring costs of their employees' health care and health insurance, absence from work, and low productivity, not to mention the costs of workers' compensation, employee turnover, accidents, and losses due to premature death of key employees. The companies that have already organized corporate prevention, health education, and fitness programs are persuaded that the benefits of the program outweigh the costs. The purpose of this book is to report how they came to that conclusion, so others can be persuaded to go and do likewise.

Chapter Two
INDUSTRY AT THE HEAD TABLE

Concern for employee health didn't come to industry all of a sudden, as the vision came to Saul on the road to Damascus. A professor of medicine at the University of Padua in 1700 wrote a treatise on "trade diseases," so medical students would learn how to protect workers in hazardous occupations. In the 19th Century the industrial revolution alerted communities and employers to work hazards and produced the first laws protecting women and children, especially, from conditions that were often oppressive. Widespread disregard for the deleterious health effects of working conditions was one of the iniquities of industry that so exercised Karl Marx, whose *Das Kapital*—among other consequences attributed to it—is thought to have been influential in the introduction of the first workers' compensation laws in Germany in 1884. At about the same time, court decisions in Britain and the United States began to formulate policies having to do with employer liability and specifying obligations to provide safe conditions of work. Not long afterward, the states undertook the long process of evolving the American workers' compensation system, with industrial medical departments as centers of responsibility.

Through the early years of this century, medical departments remained principally concerned with the prevention of industrial accidents and diseases. Then the rapid expansion of industry during and after World War II broadened the departments' interests to comprehend what came to be called occupational health, aimed at protecting and improving the health of all workers, not just those whose work exposed them to hazardous conditions and unhealthful environments. Medical department recommendations often led to improvements in lighting, ventilation, heating and air conditioning, noise control, and other elements of the environment affecting worker comfort as well as health per se. These developments were not totally unrelated to the growth in size and power of unions over the same years, but managements were nonetheless pleased to observe that improvements in the work environment were often accompanied by improvements in worker productivity.

The postwar years and ensuing decades were also a period of rapid growth for hospital and medical prepayment and insurance. Born in the depression years, Blue Cross and Blue Shield Plans, and the health insurance companies that got into the act a little later, became part of an extraordinary confluence of the separate forces of industrial expansion, the rising strength of unions and industry-wide collective bargaining, the explosive advances of medical science and consequent specialization of medical practice in hospitals, the rapid modernization and expansion of hospitals that had stood still during the years of depression and war, and the mounting awareness and expectations of a burgeoning population. These currents were all flowing in the 1950s and 1960s, and the prepayment concept provided a sort of unified field theory for the financing and delivery of medical services, broadening and accelerating the rush of events and sweeping Blue Cross and Blue Shield Plans and insurance companies along with them.

Too Much Too Soon

But as everybody now understands, the unified field theory created excesses as well as successes, and for the past several years worry about the former has virtually silenced cheering about the latter: Hospitals in some areas got overbuilt and in some cases overused, with comprehensive health insurance bargained for by unions and paid for by employers making it easy for doctors to "order the works" for their patients, while hospitals collected for services rendered under cost reimbursement contracts.

Fueled by the double-digit inflation of the late 1970s that nobody had found out how to control, the system became obtrusively expensive, and, as always happens when problems get out of hand, there was as much finger-pointing to assign blame as there was head-scratching to find answers—with everybody pointing the finger at everybody else. Hospitals were blamed for expanding too far too fast, doctors for ordering too much too soon, Blue Cross and Blue Shield Plans and insurance companies for covering too many services with too few constraints, unions for demanding benefits and corporations for providing them, government for overreaching with the new entitlement programs and then overreacting with regulations to curb them, consumers for overusing the whole system. There was some truth to all these allegations, but their endless repetition, inevitably seized and displayed in headlines and on television, tended to obscure the facts that nearly half of the towering health care cost rise was attributable to the general inflation, and the greater part of the other half represented advances in scientific medicine and its technology, resulting in improvements in care that nobody wanted to forgo. The cost of unnecessary services or abuses was a fraction that nobody could measure.

Medical necessity is in part fact that can be measured against objective standards, and in part abstraction that exists in the minds of physicians and patients. Thus one man's necessity is another man's excess, and one era's need is another's extravagance. Victor Zink, General Motors director of employee services and benefits, explained what happened. "All elements of the health care system contributed to the problem," he said at a meeting of hospital financial executives. "People wanted hospital beds, so we built them. People wanted more doctors, so we educated them. People wanted first dollar insurance coverage, so we provided it. Whether we would have done differently if we had known what we know today is immaterial. What's done is done."[1]

What's done isn't easily undone, but everybody is trying. In spite of the difficulties inherent in the nature of the services, there have been concentrated efforts to eliminate abuses and conserve health care resources, at first by government regulation, and, for the past three years, by an organized voluntary cost containment enterprise joining groups of doctors, hospitals, insurers, and business interests. The Voluntary Effort, as it is called, hasn't been wildly successful, perhaps, but it has been a factor in slowing the rate of increase in health care costs, keeping it at or near the rate of inflation in consumer prices generally—a marked improvement over the performance of preceding years. Moreover, the Voluntary Effort was successful in forestalling further restrictive legislation and regulation of health care institutions. With the advent of a new federal administration and a new Congress following the 1980 elections, it seems likely that the Voluntary Effort will continue, regulatory restraints will be relaxed instead of tightened, and legislative reform will seek to encourage competition in the provision and financing of health services and replace incentives to spend with incentives to save. There are disputes still about how this should be done and what results can be expected, but the direction is clear. Meanwhile, all the parties at interest have become involved in the wellness movement, with industry, no longer a silent participant in third-party contracts, like an absentee host who pays the bill but doesn't come to the party, now sitting at the head table and asking questions about the bill.

Coalitions for Health

In many communities, in fact, industry has taken the lead in organizing business coalitions for health that function to seek control of health care costs through areawide health planning, utilization review, promotion of health maintenance organizations, and employee fitness and health promotion programs. The coalitions, which sometimes include representatives of labor, local government, health insurers, and providers of health care services, as well as business, are being organized with the encouragement and cooperation of the Washington Business Group on Health, a national organization representing 200 major corporations concerned about health care costs and determined to do everything they can to ease the burden. In mid-1981 there were reported to be some 75 local "business coalitions for health" in operation, with more being formed and continued growth of the movement a seeming certainty. Plainly, industry is going to keep its place at the head table and keep on asking questions.

As a prudent buyer's questions should, industry's questions go to all the components of the bills it is paying for health care, and the most basic questions of all go to the point at which health care cost begins: Why do employees get sick in the first place, and what can we do to prevent it from happening? And since prevention has always been the business of the company medical department, it would be logical to assume that the initiative for a company-sponsored program of prevention, health education, and physical fitness would come from the company's medical department. One would think that the doctor or doctors and nurses who are on the premises all or most of the time, who are responsible for looking after those who get sick or hurt on the job, and who could be expected to be most familiar with the health status of the work force, would be the first to see the need for a broadly based health promotion program, to lobby for the idea, get the program going, and either run it or oversee those who do.

One would think that would be the case, but one would be wrong. Except in a very few instances, the initiative comes from somebody else. In large corporations, it has usually been the executive who is responsible for the corporate benefit program—pensions, group life insurance, and, most importantly, health insurance. It is the benefit manager, not the company doctor, who knows to the penny how much the employees' health insurance premiums were increased this year, and last year, and the year before. It is the benefit manager who is familiar from week to week with the number of employees and their dependents who are admitted to hospitals, and who knows what they were admitted for and how long they stayed. It is the benefit manager and his or her assistants who are on the telephone with the prepayment plan or insurance company, straightening out misunderstandings about claims and payments, and who furnish all the data that are needed for negotiations with the union when contract time comes around and health insurance benefits are valued chips on the bargaining table. It is the benefit manager whose eyes light up first when he or she considers how a well-rounded health promotion program can cut down the number of hospital admissions and claims and keep employees healthy and on the job; after that it is the treasurer, and the president, and the chairman, and the shareholders on the board, who get stars in their eyes when they learn about the plan.

In smaller companies it may be an executive whose responsibilities include health insurance, as well as other personnel or financial areas, who becomes interested in health promotion for employees. Often it has been the president or chief executive officer, especially if he is interested in fitness himself and has become a jogger or a believer in the Breslow plan for healthy living. In many companies, large and small, the sudden death of a young or middle-aged key executive has resulted in mandatory periodic medical examinations for all high level managers. When these examinations reveal, as they often do, that several executives are seriously overweight, or have high blood pressure or other threatening symptoms or conditions, top management may decide to institute a fitness program for managers at or above a stipulated level. A gymnasium or exercise room at corporate headquarters is set aside and equipped, and staffed with a full-time or part-time fitness director. Menus for the executive dining room or company cafeteria are overhauled to emphasize fresh fruits and vegetables and play down or eliminate fats and sweets. Smoking is restricted; weight loss scores are kept and watched jealously, jogging distances and times are matched and argued about. Fitness becomes a topic of company conversation, more sporting event than chore, and the corporation is the better for it.

In many companies of all sizes, interest in fitness begins, and ends, at the executive level. This is especially true in large, diversified corporations, where the complexities of organizing programs for employees dispersed in plants and offices all over the country, doing different kinds of work in different kinds of business, are often formidable. Even so, many of these companies have gone ahead and initiated programs, usually beginning with corporate headquarters personnel and moving on to other locations as experience shows that it is worthwhile and can be done.

It Worked at Xerox

Xerox Corporation, for example, started its first employee fitness program in 1967 at corporate headquarters in Rochester, NY, where 12,000 of the multinational company's 65,000 employees work. Today there are four Xerox fitness facilities in Rochester alone, six others in the United States, and three in Europe. And since there are Xerox employees in numerous other locations, but not enough in any of these places to justify investment in a fitness facility, the company in 1979 initiated a health management program for all employees. The program is home- instead of work place-based, and the company furnishes materials and instruction in health and fitness techniques. To conduct the health management program, George Pfeiffer, program manager, reported, 150 team leaders were trained at the company's extensive fitness facility in Leesburg, VA. The team leaders are mostly Xerox sales people, who are expected to spend from 8 to 12 hours every month making health management presentations, giving tests, and generally supervising and reinforcing the program, which begins with an employee's self-measurement test. In the first year, 8,000 Xerox employees completed the first phase of the program and received T-shirts proclaiming "I'm Taking Charge of My Life"—the title of the program. Employees also receive a "Take Charge" newsletter, each issue of which is devoted to a specific health problem such as stress, smoking, fitness, fitness for children, and others. These subjects are also featured in quarterly workshops, and support activities are planned to stimulate interest and motivate employees to accept responsibility for their own health.[2]

The goal of a similarly comprehensive program planned for a different kind of employee was described by executives of Chicago's Continental Bank. "We expect to educate and encourage our personnel to attain a lifestyle which embraces good health practices and prepares them to become knowledgeable health care consumers," they said in a background paper prepared for a Boston University conference on corporate approaches to employee health programs. In addition to screening, health education, and nutritional features, it was expected that the bank would offer seminars to reduce stress related to work and career needs, marital problems, divorce, teenage drug abuse, mental health, and child care. Another goal of the Continental program was the development of a data collection system that would provide a base for epidemiological studies to evaluate trends and program effectiveness. The planned program "is based partly on altruism but, more importantly, on the need to limit costs due to absenteeism, illness, hospitalization, and downright job maladjustment," it was explained. Critics might point out problems for such a comprehensive plan, the authors acknowledged, "First, the limited impact that can be expected from job-site intervention in a primarily urban, 9,000-person work force, made up of 60 percent women with a ratio of black to white at about one to two and an age range of late teens to middle seventies; second, the difficulty of inducing permanent changes in behavior patterns. It is a common observation that the overweight worker goes on the wagon January 1 or gives up desserts for Lent and drops 15 or 20 pounds. By the time the subsequent winter has arrived, however, he is back up to the original level of obesity. Our hopes are based on the belief that the firm, persistent pressure of education and peer performance may help us in this regard."[3]

There may also be some problems with employees, it was suggested. Some might resent or resist the implications of paternalism, for example, and "the sad fact is that in certain instances healthy eating is more expensive than junk foods. It is also true that many individuals will exercise only if they are given time off from their jobs. We recognize these facts but they do not deter us. Again, our educational efforts are most carefully planned, our suggestions as tactful as possible."

To help control the escalating costs of health care and improve the level of health of its employees, International Business Machines has initiated a new health education program called "A Plan for Life." Operating in all cities where IBM has operations, the health promotion program utilizes community organizations and qualified instructors to offer both mini and comprehensive courses on several subjects related to good health. The courses are offered before and after working hours.

Regular and part-time IBM employees and their spouses, as well as retirees and their spouses, are eligible to enroll in the program. Employees' dependent children 15 years and over also are eligible for certain courses.

One and a half hour mini courses are offered in health and nutrition and risk factors management. Comprehensive courses, conducted over a period of two to 12 weeks, are on topics such as exercise, smoking cessation, stress management, weight management, healthy back, first aid, and CPR.

"Our decision to open these courses to employees' families

and retirees was based on the fact that two-thirds of the medical costs incurred under our health insurance programs come from the family and the retiree, not from the individual employee," said Robert Beck, IBM director of benefits and personnel services. "Just to focus on the employee only would not be as cost effective."

According to Beck, the decision to use community facilities and resources for the IBM health promotion program "permits a low cost, rapid start-up across the country, rather than building our own facilities or hiring people. We also wanted to minimize the effect on business operations," Beck said in an interview. He stated that the initial response to the courses was strong and that in most cities where the courses were initiated, IBM was able to offer all the courses because of the heavy sign-up. Where there wasn't sufficient sign-up, those eligible for the program were allowed to take comparable courses in the community and get tuition assistance. Approximately 25 percent of the participants are spouses.

The position held by Beck at IBM is a part of the corporation's new approach to the overall health of its employees and its total health care costs. As IBM director of benefits and personnel services, Beck's responsibilities include all employee benefit programs, health and safety programs—including medical services—and other employee services. "This enables us to take a coordinated, efficient approach to our overall health profile and costs and gives us new opportunities for cost savings and improvements in employee health and productivity," said Beck.

If new opportunities for cost savings are good for a large bank and even larger multinational corporations, they are also good for the company with as few as 50 or 100 workers, for whom health education and physical fitness programs are equally important, and equally effective. There is nothing wrong with a company fitness program that consists entirely of an occasional pay-envelope stuffer on nutrition and a memorandum from the boss suggesting that it would be a good idea for interested employees to join the YMCA or YWCA and sign up for exercise classes. There's nothing wrong with that, but it isn't a company fitness program, and it isn't going to accomplish much of anything. The small employer has as much at stake, and as much to gain by an organized effort to improve and maintain the health of employees, as General Motors has. In both cases, there's no sense in doing it wrong, and in both cases it costs something to do it right, and in both cases the cost should be regarded as an investment. Like an investment in insurance, it's a form of protection against loss.

The small company may not be paying for employee health care or health insurance, as most large corporations do today, but the small company has exactly the same interest in the productivity of the work force. Days lost to illness and work time or work quality lost to fatigue or boredom or inattention are costly to any enterprise; health education and fitness programs planned to meet the carefully assessed need are tested methods of minimizing these losses. The question is not whether a company, of whatever size, can afford to have such a program. The question is whether it can afford not to.

What About Results?

It isn't an easy question to answer, because there are complexities to evaluating the results of these programs in dollars and cents. The processes of evaluation will be examined in a later chapter, but meanwhile there is an abundance of persuasive evidence suggesting that, if anything, the pay-off on fitness programs in industry is greater than anybody had supposed could be the case. A professor from the Soviet Union, for example, reported on an exercise regimen for one group of workers; careful observations were made and records kept of its experience compared to that of a similar group that had no program. After an extended period of observation, the work output of the exercise group exceeded that of the nonparticipants by 10 to 15 percent; the former were judged to be more disciplined workers than the latter, consulted doctors only one-fourth as many times, and had a sickness rate half that of the nonexercisers.[4] In a Goodyear factory in Sweden absenteeism was cut in half following introduction of a comprehensive fitness program.[5] In another widely reported study, the National Aeronautics and Space Administration initiated an exericse program for 260 middle-aged men and reported that half the regular participants in the three-times-a-week program had better job performance and better attitudes toward work. Only 12 percent in a group that took part "now and then" had similar job and attitude improvements. All the regular participants said they felt better; 89 percent reported improved stamina, 40 percent sounder sleep, and 60 percent lost weight. Half said they paid more attention to their diets, and many either quit or cut down on smoking. And most significantly, NASA reported a consistent and positive relationship between these perceived benefits of the program and the results of medical examinations.[6]

Work output up 15 percent? One-fourth the number of doctor visits? Half the absenteeism? Half the sickness rate? Better job performance? Better attitudes? What are we waiting for? Let's get going!

Now wait a minute. Results like that aren't achieved overnight, and they don't come easily. It requires a positive commitment, with the understanding that results aren't going to show up right away. The Goodyear factory mentioned here is in Sweden, and there, as in many European countries, "exercise breaks" and comprehensive fitness programs have been an accepted practice for years, so employee participation comes naturally. Certainly, Americans are a lot readier to join in any such program today than they would have been 10 years ago, when the occasional jogger was seen as some kind of nut; even so, motivating employees to believe and take part has to be built into every program from the beginning. And the motivation effort has to be sustained. Everybody who has had to diet to lose weight, or who has quit after years of smoking, knows how easy it is to get discouraged and think "the hell with it." So the employer or fitness director or staff member or whoever is in charge *has* to believe, and *has* to be part coach and part cheerleader, and above all *has* to be committed.

With absenteeism costing American industry an estimated $20 billion a year, not to mention the amounts spent for sick pay and health insurance and time lost on the job that can't possibly be measured, the commitment to a company program of prevention and health education and physical fitness should be worth whatever it may cost. It doesn't have to cost a fortune; but at any size it will cost something for space and equipment and educational materials and somebody's time, and these details will be considered in later chapters. So an employer who isn't willing to spend something before he can expect to get back his share of that $20 billion would do better to forget it. He might just as well try to jog on one leg.

Chapter Three
THE WAY TO GET STARTED

When a decisionmaker becomes convinced that the company ought to consider a health or fitness program for employees—because he or she has heard or read something or talked to somebody about it, or is just aware that a lot of others seem to be doing it—the first question is likely to be, "What kind of health promotion program should we have?"

The only sensible answer is another question: "What kind of company have you got?" Obviously, the question relates partly to company size: What's good for General Motors is not necessarily good for everybody, even though the principles and goals are the same. For one thing, company size is an important determinant of the kind of facilities that may be available. A multiunit corporation whose plants and offices are mostly in spacious industrial parks or suburban sites will plainly have options that aren't available to the company housed in a single structure on a crowded city block. One firm may have a warehouse—part of which can be converted to gymnasium use—and another a spread-out factory whose roof is just the place to build a running track. Most smaller companies will probably have a hard time finding or making space for a minimum exercise room and accompanying shower and toilet facilities, but it's surprising what can be done when there's a will to get it done.

An executive who has become a fitness enthusiast tells of calling on a friend at his office on the top floor of his company's headquarters building, overlooking a beautiful, expansive park. "What a perfect place for our employees to jog, run, exercise—whatever they want to do!" the friend said as they stood looking down at the park. "But they just don't take advantage of it. They don't use it at all."

"It never occurred to him that the reason they wouldn't use it was that there was no place for them to keep running clothes and change and shower," the executive said later. "Nobody wants to go out and jog or run or play tennis or exercise and then go back to the office or desk and sit there steaming. Somebody has to think of things like that."

Something for Everybody

The question about what kind of program to have relates to the nature as well as to the number of employees. Age and sex distributions are important; a predominantly young group may respond to a different kind of exercise regimen than older workers would, for example, and women may be more interested in diet and weight control than men would be. In fact, they are. In a recent survey of government employees in Canada, the primary reason given by women for participation in fitness programs was to lose weight and improve their figures, whereas for men the most important goal was to get themselves "back in shape" and "feel the way I used to." Relief of tension was also mentioned frequently by men.[1] That doesn't mean the program should be designed for one group and not the other, but since maximum participation is obviously desirable, age and sex distribution may help determine the kinds of motivational and educational materials that are likely to be most effective. The type or types of work also need to be considered. Workers who sit at desks all day have needs for exercise and relaxation that are different from those of workers who stand at machines all day, or drive trucks, or lift boxes. Most companies of any size have all kinds, but, again, the mixture may be important to program planners. So is the level of education, or the mix of levels, which is critical in determining what workers will understand and respond to. In certain parts of the country, too, and in parts of most big cities, ethnic background will be an important determinant of interests and attitudes affecting participation, and perhaps program content.

Companies having more than a handful of employees generally provide some kind of dining facilities—lunchroom, cafeteria, coffee shop, snack bar, or at least a room or area of vending machines dispensing foods. The top selling items at most of these places, invariably, are hamburgers and hot dogs, chili, potato chips, sweet rolls, pastries, candy, ice cream bars. These foods can't be ruled out altogether when a fitness program is initiated; too many employees would be disaffected, and the purpose is to prevent turnover, not encourage it. But food policy should be reviewed to make certain every station offers healthful options. If a feeding contractor furnishes foods and food service, as is commonly done, the contract should be revised as soon as possible to be consistent with fitness goals. In any case, cafeterias and snack bars are effective sites for diet posters, calorie charts, and other pertinent educational displays. And somebody has to make a judgment call about smoking. It doesn't make any sense to have a cigarette dispensing machine in the lunchroom that is used after hours for smoking cessation classes, and limiting smoking to restricted areas is now accepted public policy and common corporate practice. But how far to go with restrictions is a balancing act that every company has to perform for itself—with the certain knowledge that any policy or rule that may be adopted is certain to offend somebody.

Familiarity with the company's personnel records as these relate to health is another essential for planning the health promotion program. Turnover rates, absentee rates, sickness rates—days lost on account of illness, by departments and types of work if these figures are available, are all worth studying. The personnel or medical department, if there is one, should also have records showing the types of illness responsible for time lost. These figures will provide clues to the opportunities for improvement, and perhaps suggest what kinds of education program might be most useful. In any case, the figures require interpretation by an experienced observer. If a particular department or job type is characterized by a high absentee rate with repeated designations such as headache, backache, colds, fatigue, digestive upset, and other synoptic causes, for example, it may be that these people are subjected to unusual problems of stress or morale. A management interpretation might be that they are faking or malingering and what is needed is discipline; maybe a few firings would straighten them out. But modern health promotion practice would suggest classes in stress management or relaxation technique, as well as diet and exercise, with a resulting benefit to turnover and absentee rates, instead of the accelerated rates that firings and added stresses would be likely to produce.

Records Provide the Base

Other company experience needs to be examined to complete the assessment of employees' health status. Medical department records will show that on-the-job accidents and illnesses have been handled by company doctors and nurses, with referrals to employees' physicians. Personnel or benefit departments will produce information on health insurance claims and costs, including whether the employee, spouse, or dependent is using the benefit. While these records can provide some guidance for planning the health promotion program, they will be very useful in furnishing a baseline from which progress can be measured over time. Planners should avoid any expectation that health insurance costs are going to be influenced right away. These costs will continue to rise unless there is a dramatic improvement in the general inflation rate. But it can be expected that the rate of increase in the cost of health insurance will be slowed by a continued comprehensive health promotion program.

From a review of all these pertinent records, an experienced health promotion or fitness director or consultant can make a reasonable assessment of the work force health status. If a consultant isn't available, a personnel or medical department or health insurance executive at least will know or can readily find out if such important indicators as the absentee rate and days lost on account of illness are high, low, or about average for similar operations; in any event, a health promotion program will be useful, but a high absentee or sickness rate should help persuade management that the investment will be worthwhile. For purposes of program planning, the company's health status record needs to be supplemented by an appraisal of its resources. We have already seen that the available indoor and outdoor space and facilities, and especially the food service facilities, are critical factors. And certainly the existence, attitudes, and resources of company medical, personnel, and benefit departments are equally important.

In addition, the person or persons responsible for planning the health promotion program should know the disposition of top management toward such a proposal. What does the chief executive officer think about it? Is he or she generally interested in employees and convinced that an investment in their health in the long run will produce improved performance, stability, and morale? Or has the health promotion idea been sold by an enthusiastic executive or group, to the point where the chief has consented to give it a whirl but is likely to lose interest or become impatient if positive results can't be demonstrated within a few months? The chief executive's attitude, along with the availability and convenience of facilities, may determine whether a comprehensive program requiring a major investment can be initiated right away or whether it would be wiser to start with a comparatively modest proposal for health education and a pilot exercise program, and build interest and participation toward an expanded investment at a later time.

A key group whose feelings must be considered in arriving at an answer to the same question is the supervisory personnel. If there are a few department heads or line supervisors who are known to be joggers or swimmers or "healthniks," their help can be enlisted to take the temperature of their peers and determine whether some kind of preliminary or preparatory educational or promotional effort for supervisors would be advisable. If such pathfinders can't be identified, a questionnaire survey of supervisors may be needed. To a substantial degree, the enthusiasm and cooperation of this group will enhance participation and ultimate success of the whole program, and supervisory indifference or opposition will diminish participation and might even destroy the program. The same thing can be said about union representatives in plants and offices that are unionized. Most unions are in favor of programs that can be expected to benefit their members, and, in fact, in many cases unions have initiated health education and fitness plans. Nevertheless, a union that has been involved in a dispute with management may automatically oppose anything that is presented by management, unless union representatives have been consulted in advance. The point is that the union attitude should be known, and union support enlisted, before final plans are made.

Employee Opinion Survey

If union or supervisory attitudes are found to be indifferent or antagonistic, as may be the case if supervisors see the proposed program as invading work time and creating added responsibilities for themselves, keeping track of employees and keeping them on the job, and if educational sessions for supervisors fail to ensure a reasonable degree of cooperation, it may be wise to consider a company-wide survey of employee opinions about health, health education, and fitness. This should be done with the cooperation of the interested departments—personnel, medical, benefits, and above all it should be kept simple. A detailed questionnaire about health might cause uneasiness and create more questions than it would answer. The following questions, or something like these, should be enough:

> Do you consider that you are generally healthy? In good shape physically? Could be better?
>
> Would you like to have more information about:
> Symptoms of common diseases?
> When to see the doctor?
> Healthy diets?
> Effects of smoking or drinking?
> Healthy living habits?
> Other things related to health? What?
>
> Do you get regular exercise at least two or three times a week? What kind?
>
> If not, would you if it were easier and more convenient for you to do?

The questionnaire should be anonymous, but it might be smart to have the respondents identify their departments. Then if it seemed sensible to start small and build participation, the group most interested in having health information and taking part in a fitness program would be a logical place to begin.

Because communications will have an obvious effect on the whole health education and fitness program, as well as on the method used to introduce it and get it started, the company's system for communicating with employees should be studied as an important feature of the resource appraisal. Is there a company newsletter, bulletin, magazine, or house organ of some kind that can be used for announcements or reports about the program? How often is it published? How distributed? Have studies been made to determine how widely and thoroughly it is read? Would it lend itself to the inclusion of some health information?

Are there bulletin boards where notices may be posted and wall space where posters could be displayed? How often are these methods used? How effective are they judged to be? What is company policy and practice regarding employee meetings? Is there an auditorium or cafeteria that would accommodate all employees? Have such meetings been held? What for? How have night shifts been handled on these occasions? Are departmental meetings held? Where? Could such meetings be held during working hours, or only after hours? Is there a loud-speaker system or systems that can be used for brief announcements and reports? How is it usually used? How effective is it? Are there meeting rooms, conference rooms, or offices that can be used for small groups, such as smoking cessation or weight control classes? When would these be available? What other methods have been used to inform employees about company policies and activities?

For the health promotion program, certainly some other than these routine communication channels will be used. Special health education materials will be distributed, particularly, and detailed instructions are required for people with special interests or problems. In a large company there may readily be need for a separate health education and fitness publication of some sort. But the regular company publications and communications systems are important for backup use, announcements and reports. And there is no substitute, in health education as in other communications, for the face-to-face interaction provided in meetings, with plenty of time allowed for employees to ask questions, express opinions, and report their own experiences. In fact, meetings of small groups of employees who are participating in the same activities and pursuing the same goals can build the kind of camaraderie that boosts morale—and of itself is a form of relaxation that relieves stress and contributes to health and fitness.

Small companies whose facilities are so limited that it would be impractical to try to accommodate either employee or discussion group meetings or any kind of exercise program on company premises may still have useful health promotion programs. The task is to identify the community or neighborhood institutions, organizations, and locations that can most conveniently be adapted for health program purposes. The most likely candidates are schools, colleges, YMCAs, hospitals, parks, churches, and such common meeting sites as motels, hotels, and clubs or organizations that have headquarters and meeting facilities not constantly in use. Someone, or perhaps an employee committee, familiar with the requirements of the planned activities, must canvass all the possibilities within reasonable reach and select the one or ones that provide the best combination of location, cost, and suitability for the purpose. It is unlikely that there could be a company or group of employees anywhere that is really determined to have a fitness program and unable to find a place and way to make it work. After making an assessment of company needs and resources, most firms have decided to seek outside help in getting an employee health promotion program started. Several kinds and sources of such help are described in the remaining chapters of this book, following a discussion of some of the common health problems that are likely to be found in any group.

Chapter Four
COMMON PROBLEMS

In any group of men and women who consider themselves healthy in the sense of the word most commonly intended —that is, they are not sick — a substantial number have conditions that are subtracting something from good health. In most cases these men and women are unaware that there is anything the matter with them. They feel all right, as they would express it. Yet their conditions mean that they are living and working at less than full effectiveness. They are predisposed to illness; they lose more than the average number of days at work, and, most seriously, if their conditions go undetected and untreated, there is every likelihood that they will cause serious or disabling illnesses or breakdowns of some kind.

The most common of these conditions, nearly all of which can be discovered and either eliminated or substantially improved in a comprehensive company health promotion program, are obesity, hypertension or high blood pressure, alcoholism, stress, and mental or emotional upset. Because some or all of these will be encountered in any group of employees, they are well worth considering when the program is in the planning stage, so the plans can include strategies for dealing with them.

Obesity

The Public Health Service in the U.S. Department of Health and Human Services has estimated that 30 percent of all men over age 30 are at least 10 percent over the best weight for their height and body types, and another 30 percent are at least 20 percent overweight — the point at which the condition is judged to become threatening. Among women, the percentage in the over-30 age group is somewhat lower than it is for men, but starting at age 40 substantially higher than the figure for men.[1] Moreover, studies from the same source indicate that for both men and women the problem has been worsening, a circumstance that is attributed to changing dietary habits. Shortcuts in food preparation methods have us all eating more and more processed foods, and at the same time we are taking in more calories we are also becoming more and more sedentary, requiring fewer calories for maintenance. In other words, we are eating more and using less, and the difference is in our figures—ours and the Public Health Service's.

There are other causes contributing to obesity. Thus there may be metabolic disorders, disturbances of the mechanism by which the body absorbs fats and carbohydrates, and there may also be genetic causes, or inherited factors. But physiologists and nutritionists agree that when obesity runs in the family, as fat people often like to say, excusing themselves, it isn't as likely to be an inherited metabolic or glandular defect as it is to be a familial pattern of overfeeding imposed by obese parents on their children, often reflecting cultural dispositions that have persisted for generations. In any case, the total of all the other causes accounts for only a small fraction—some say as little as 5 percent—of obesity in the United States today. Fat comes from food, and too much of one means too much of the other.

Wherever it comes from, fat is dangerous. Fat people don't live as long as thin people do, because they are fat. In an interesting study of men who were charged higher life insurance premiums solely because of overweight, the mortality rate of those who were moderately overweight was 142 percent of the rate for those who weren't considered poor risks. The mortality rate for markedly overweight men was 179 percent of the standard risks. "The common partners of obesity are increased blood cholesterol, blood pressure, and blood glucose levels," said Beverly Winikoff, M.D., of the Rockefeller Foundation. "Through these effects, obesity is an important risk factor for coronary disease, hypertensive illness, stroke, and diabetes. Reduction in obesity would undoubtedly lessen the incidence of these diseases."[2] Obese people also are inclined to have more respiratory difficulties, and more respiratory infections. "The more weight there is in the chest wall, the greater the work of breathing," said the Public Health Service report. Obese women are more likely to have menstrual abnormalities, more disorders associated with pregnancy, and more difficulties and complications during delivery, the report also said.

A Complex Problem

Fat people tend to have more psychological disturbances than are seen among others, but nobody really knows whether they are disturbed because they are fat or fat because they are disturbed. "Behavioral consequences of an obese appearance, ranging from shyness and withdrawal to overly bold self-assertion, may be rooted in neuroses and psychoses," says one authority. "Psychological disturbances are often responsible for the overeating that causes many patients to become and remain obese," says another. "No single concept provides the explanation for either the cause, the treatment, or the psychic manifestations of a problem that is so complex," says the Public Health Service.

If the problem of obesity is so complex that doctors don't know the answers, how can it be dealt with successfully in a company-sponsored health promotion program?

Obviously, the problem can't be dealt with successfully for every obese person in a company program. But the fact is that it *is* working for many in these programs, and the reason is that for many of these obese people the problem isn't really all that complicated. They don't want to be fat; they don't like the way they look. If they know about the risk of coronary disease, or stroke, or diabetes,or diminished life expectancy, they try not to think about it. They try every new diet that comes along. They lose some weight, then they lose interest and get fat again. One of the most successful of all the weight control programs, Weight Watchers, has found that most of the people who take part gain back the lost weight; many of the people in these programs are repeaters. Some are repeated repeaters.

But a company health promotion program provides something more than a few weeks or months of intense dieting and then "goodbye and good luck." The company program provides the powerful motivation of peer pressure. That's what makes one-time programs work, but when the program is over the peers disappear, and so does the motivation. The company program, in contrast, has continuity in its education, fitness, and support activities. The peer pressure motivation is lasting, and it is based on more than a single common goal; only family ties are stronger than work ties. The executives who make a game of fitness have their counterparts in the office and shop; it works best when they're all in it together.

Hypertension

A renowned authority on hypertension, Jeremiah Stamler, M.D., declared not long ago that, "High blood pressure remains an inadequately solved mass adult disease problem in the United States and in almost every country in the world. It is a key one among the adult health problems in this country because of its mass prevalence and its impact on sickness, disability, and death in the prime of life as well as among the elderly. It exists throughout the population, in all areas of the country, in all socioeconomic and ethnic strata, with age-specific rates —higher in men than women until the 40s or 50s, consistently higher in blacks than whites. It is often present for years without symptoms. It is important because it is at all ages powerfully related to risk of premature major catastrophes—sudden death, heart attack, congestive heart failure, stroke, and kidney failure."[3]

Dr. Stamler, who is Professor and Chairman of the Department of Community Health and Preventive Medicine at Northwestern University Medical School in Chicago, deplores use of the words "mild" and "benign" to describe the less severe forms of hypertension, as many doctors and others commonly do. "The words are misnomers and anachronisms," he said. "They came into the medical vocabulary decades ago as tools to distinguish less severe hypertension from malignant, accelerated hypertension. While they were possibly useful at that time, they have now become very dangerous words, because they reinforce a view in medicine that the approach toward the care of people with hypertension is not to treat them. All levels of high blood pressure need treatment." For some of these, observation should be sufficient.

Heredity has long been recognized as a factor among the risks of high blood pressure, but its importance has been debated, and while it is agreed that susceptibility to the condition may be inherited, the environmental causes are the most important. To the extent that it is also largely environmental, obesity is one of these risks. Hypertension occurs more frequently among obese people than it does among those who are thin; thin young people who gain weight in middle age develop hypertension more often than those who stay thin; obese people with high blood pressure may moderate the condition by dieting and losing weight. The precise mechanisms that result in these phenomena are not understood, but the statistics are unarguable. There are also striking differences in the rate of occurrence of hypertension in different parts of the country, considered to be the result of differing soil compositions reflected in food and fluid intake.

Research in nutritional factors has revealed that hypertension is rare in populations with little salt in their diets; in fact, the relationship of hypertension and salt intake is so strong that low-salt or salt-free diets are a basic approach in treating the disease. There is also a known connection between hypertension and alcohol. Heavy drinkers—three to five drinks a day—are more likely to be hypertensive than others are. "Data on this are now available from at least four epidemiologic studies," Dr. Stamler reported. "Regular consumption at this or higher levels is associated with significantly higher blood pressures, and this relationship remains significant when other factors are controlled—for example, overweight. That is, it is not simply due to extra calories and resultant obesity from alcohol." Limited data from two investigations of white males have also suggested that cigarette smoking may be independently related to the risk of developing hypertension, Dr. Stamler added.

Finally, some high blood pressure has been associated with psychologic states. Tension, stress, anger, anxiety, fear—all are associated with elevated blood pressures. Thus patients visiting doctors' offices with anxieties about health problems, or simply nervousness generated by the occasion when they are reporting, in the absence of any problem, for periodic examination, often show elevated pressures if the determinations are made at the beginning of the visit, and sharply reduced pressures when the determinations are repeated later, after the initial anxiety has subsided. But when these psychologic states are long continued and tend to become chronic, high blood pressure is likely to follow suit.

Epidemiologists estimate that there are 30 to 35 million people with high blood pressure in the United States, and some say the number is even greater — one out of every five people in the population, or more than 40 million. Whatever the number, epidemiologic studies have suggested that only about half of those with high blood pressure are aware that they have the condition. Thus the principal thrust of the public health effort in hypertension for the past several years has been in public education aimed at getting people to go to public health clinics, hospitals, or doctors' offices to have their blood pressures checked. If the epidemiologists are right, half of those who go should then be treated to bring the condition under control and keep it there. But the Public Health Service and state and local public health departments can't reach everybody with their educational messages, so the greater part of the task remains to be done. And identification of unsuspecting men and women with elevated blood pressure is among the important goals of workplace health promotion programs.

Finding the Risks

Just about everybody does know that the procedure for measuring blood pressure is simple and painless: The familiar cuff is wrapped around the upper arm; a few quick injections of air are pumped into the cuff to tighten it — and the pressure against the arterial walls is then measured in millimeters of mercury visible on the sphygmomanometer scale, first when the heart is contracting to force blood through the arterial circulation, and again when the heart cavities are expanding and filling up. Thus the two levels of measurement, expressed, for example, as 120/60 — the first in millimeters of pressure during systole, or contraction, the second in millimeters of pressure during dilation or diastole. The 120/60 figures are those most doctors would consider "healthy" blood pressure, but the term has to be used cautiously, because normal pressures vary widely among individuals. A reading of 100/60 is normal for some persons, and 140/90 is normal for others. The pressure in the latter case, however, is elevated enough to make the physician, if not the patient, a little nervous, and the physician is likely to recommend cutting down on salt and high cholesterol foods. "Don't add salt to anything—there's enough that's already been added in the cooking, or in the food itself," the physician may advise. "Eat only the lean portions of meats, and try skim milk instead of whole milk, and margarine instead of butter. And see me again in a couple of months."

Controlling diet, especially salt intake, is where treatment of hypertension begins. For many people with moderately elevated blood pressure that's all it takes. But it doesn't end there, because it has to be continued, and that isn't easy to do. People who feel perfectly well can easily persuade themselves that just a little added salt won't hurt. And the bacon doesn't have all that much fat. And the children can't *stand* skim milk; besides, it's a nuisance to have to keep both kinds. These things can't really make that much difference, can they?

The central problem in the control of hypertension can be summed up in one word: compliance. It's a tough problem. A survey among city clinic patients who dropped out of hypertension treatment found that the reason was "not because they were uneducated, or because they didn't care about their health, or because they couldn't afford the medication," the survey report said. "These patients abandoned the clinic because they were sent from one room to another, left waiting, then examined by a different doctor on each visit." It had been his experience, the physician who headed the survey said, that patients won't remain under medical care and on medication unless they are properly motivated, and a good doctor-patient relationship is essential to motivation. But "physicians are not interested in, nor do they really have the time to follow, apparently healthy patients," he added, "particularly patients who do not have any symptoms. Physicians are crisis oriented and are more interested in relief of pain, and in solving acute problems. The more specialized the physician's training, the less interested he is in asymptomatic, 'dull' patients. The expertise of the cardiologist, for example, has centered around unraveling complicated diagnostic problems or treating the patient in the emergency situation. The long-term, routine care of the hypertensive patient offers no challenge."[4]

The challenge is to help those on treatment to achieve long-term blood pressure control. A work-site health promotion program, in combination with physician care, is one way to get that result. Peer group influences, counseling by occupational health staff, and articles in plant, office, and union publications are all effective ways to educate employees about high blood pressure. Recent evidence has shown that comprehensive work-site blood pressure control programs are successful. These programs include educating all employees about high blood pressure, screening employees to identify possible cases of high blood pressure, refering those with high blood pressure readings to their physicians for treatment, and, most importantly, long-term monitoring and follow-up of diagnosed hypertensives at the work site. Programs with all four of these elements have helped 75 to 85 percent of hypertensive employees get the condition under control, and keep it there.

Keeping it there is the trick, and that is what makes the work site so effective for health promotion, especially for employees who are forgetful about compliance with doctor's orders. This is where peer pressure counts, and even in small companies there are certain to be others in the same situation, with the same dietary restrictions, or medication regimens, or both. Health records themselves are confidential, but if the program director schedules and announces a lecture or class or discussion group on diet, for example, or on cardiovascular diseases, or perhaps in larger companies on hypertension specifically, those with the same problems inevitably become known to one another. Soon they will be trading reports of their abstinences and their lapses, their triumphs and their hardships — and the powerful motivating energy of peer support will be released and begin to make itself felt. There will be a few dropouts and backsliders, inevitably, but over time there will also be fewer absences from work, and fewer heart attacks, and fewer messages of condolence to bereaved families. That's what fitness programs are for.

Alcoholism

Alcohol abuse and alcoholism —the latter is a form of the former —cost the nation an estimated $40 billion in 1979, a formidable amount that has probably become $50 billion or more a year by this time, what with inflation, the rising use of alcohol among young people, particularly, and the comparatively modest rates of cure. The $40 billion estimate was made by the National Institute on Alcohol Abuse and Alcoholism (NIAAA), a component of the National Institutes of Health, which is part of the federal Department of Health and Human Services. According to NIAAA, $19 billion of the total was in production lost to business and industry through workers absent on account of drinking; $12 billion was for health and medical services; $7 billion the cost of motor vehicle accidents attributable to alcohol, and $1 billion each for criminal justice and alcohol research programs.

Actually, the cost to business and industry may be much higher than NIAAA has supposed. You can count workers who aren't there, but how can you measure the losses caused by those who show up but can't really function? Or those whose attention and productivity are impaired by worry about an alcoholic spouse or other relative? The warnings about defects in cars that come off the assembly line on Mondays are an example. Most employers have examples of their own. And what about the alcoholics who leave lighted cigarettes lying around to start fires? NIAAA guesses that fires caused by drinkers cost a half billion dollars a year.

Control Is the Key

If that sounds like a lot of fires, it's because there are a lot of drinkers—heavy drinkers, that is—alcoholics and "problem drinkers." The difference has been defined by the American Medical Association: "Alcoholism is an illness characterized by preoccupation with alcohol and loss of control over its consumption such as to lead usually to intoxication if drinking is begun; by chronicity; by progression; and by tendency toward relapse. It is typically associated with physical disability and impaired emotional, occupational, and/or social adjustments as a direct consequence of persistent and excessive use."[5] NIAAA makes the distinction between those suffering from alcoholism and problem drinkers. The latter, it says, are "those who drink to such an extent that self-control is impaired and normal social behavior patterns are disrupted." Plainly, there is no hard and fast line between the alcoholic and the problem drinker; there is no moment when the problem drinker becomes an alcoholic, but the unfortunate fact is that unless he or she can be identified and treated, the problem drinker is going to become an alcoholic.

Of course, that doesn't mean that everybody who drinks—half the population, give or take a percent—is poised at the top of the slide waiting for the push that will start the skid into alcoholism. The greater number are "social drinkers." They take a drink now and then, or maybe every day, if they feel like it, maybe on occasion more than one, or even one too many. But they can also stop. It's no big deal. They may go for days or weeks or months without a drink, or they may take a cocktail or a nightcap, or both, routinely. But the important thing is that they are in control; they decide when and how much, whereas the problem drinker's control is impaired, and the alcoholic's is lost.

These characteristics of drinkers make it impossible to count the numbers of people in drinking categories with any degree of accuracy, so what NIAAA does instead is make surveys of alcohol consumption on which assumptions about drinking behavior can be based. Forty-two percent of the population age 13 and over either abstain or take less than one drink a month, and it seems reasonable to call these people nondrinkers. Thirty-one percent consume an average of ¼ ounce of alcohol or less a day. That would be a can of beer or a glass of wine every other day, say, or a stiff drink once a week. These people we can safely call social drinkers—and not too social, at that. NIAAA calls them light drinkers. Eighteen percent are what NIAAA designates as moderate drinkers; they average ¼ ounce to 1 ounce a day of straight alcohol—up to 2 ounces of 100 proof distilled spirits.[6] There could be some problem drinkers here, but they'd have to go on the wagon a lot to keep the average that low. So the problem drinkers and alcoholics are almost all found among those designated as heavy drinkers—two or three drinks a day or more, enough to addle the brain on occasion and keep the liver busy breaking down and disposing of the alcohol. Not all heavy drinkers are alcoholics or problem drinkers, as we have seen, but there is a degree of dependency here, at least, and loss of control is either present or threatening. This is where the problems are—eight or nine million problems—and this is where the $50 billion a year is being spent.

Moreover, the alcohol problem is compounded beyond the measurable disabilities and losses by the fact that alcoholics are at greater risk of cardiovascular, gastrointestinal, kidney, liver, and pulmonary diseases. The death rate for alcoholics is several times higher than the rate in the general population. Alcohol is associated with a large percentage of accidental deaths and injuries and criminal offenses. Most hidden, and possibly most expensive for society generally and not without serious losses for employers, are the social problems of which alcohol is either a contributing or a principal cause—divorces, separations, and broken families; child abuse, juvenile delinquency, crime.

Confronted with these social costs of alcoholism, it has been hard for society to accept the fact that it is a disease and not itself a crime or moral failure of some kind. And in view of the direct cost of alcoholism to employers it has been hard for them to make the same distinction. But this much has been accomplished. The society today generally understands that the alcoholic is a problem for the doctor and the medical institution, not just the police and the local judge or magistrate. Industry's acceptance is evident from the fact that there are now some 4,000 companies and unions offering assistance for employees with drinking problems.

As recently as 10 years ago it was still common practice in industry for the problem drinker to be given "one more chance" until his absences and trespasses completely outran the boss's patience; then he would be fired—and more often than not speeded on his way to disabling alcoholism. In most large corporations and many smaller companies today a staff physician or psychologist or counselor makes it possible for the problem drinker to be identified before the repetitive pattern of behavior has become obtrusively evident, and usually before the problem drinker is willing to acknowledge that he or she needs help—a point that has to be reached before he or she can be helped successfully. But the openness and ready availability of a company assistance program remove much of the need for secrecy and pretense and allow the drinker's self-awareness to rise to the surface so he can face himself and his problem and accept the counseling that is offered.

Treatment of problem drinkers and alcoholics may take any of many different courses, depending on the nature of the individual, the severity of the disease, and the resources available. In every case and at every level, however, two of the important components of treatment, in addition to the patient's recognition of the condition and acceptance of the need for treatment, are education and peer support. And for these critical elements the company-sponsored program has many advantages. Education is not just posters and pamphlets but the more effective and lasting face-to-face interaction in classes and individual counseling, emphasizing not only the physical, mental, and emotional damage inflicted by alcohol excesses but also the central fact that the problem drinker or alcoholic, unlike the social drinker, is unable to "take it or leave it alone." For reasons that are not fully understood but clearly have nothing to do with fault or blame or moral character, the alcoholic must leave it alone, period.

What Counts Is Peer Support

Once that concept is understood—and it isn't easy; some alcoholics never accept it, and for some it keeps slipping out of grasp—peer support provides the reinforcement that is the key element of successful treatment. In many if not most programs for alcoholics, peer support is provided as long as the patient is at an institution or visiting a clinic or attending classes. But then he or she returns to the real world of home and work, and all the normal stresses of living and working are still there—the stresses that build up and cause the tensions that trigger the impulse to seek relief by taking a drink. In Alcoholics Anonymous, certainly the best known and most successful of the self-help programs of its kind, peer support is on call, but the alcoholic individual must recognize the moment and have the will and the initiative to make the call instead of taking the drink.

In the company setting where assistance for problem drinkers and alcoholics is part of a broad health promotion and fitness program, peer support is endemic to the situation. Everybody is helping everybody else every day; the alcoholic under treatment and seeking help is no different from the worker who is obese, or hypertensive, or diabetic, or simply trying to keep fit. The social stigma and the feelings of shame and self-abasement it engenders are largely eliminated, and the downward spiral of drinking that leads to unacceptable behavior that leads to self-recrimination that leads to drinking is interrupted or never gets started.

Of course, the existence of a fitness program, including education and treatment for alcohol abuse, doesn't completely change the character of the work place. An alcoholic worker who has a boss or supervisor who is a martinet and thinks treatment of alcoholics is useless coddling—they ought to straighten themselves out or get fired, and the sooner the better—isn't necessarily going to be relieved of the kind of pressures that attitude produces, and there's a lot of it around still. But it is a vanishing phenomenon; a society that puts drunks in hospitals instead of jails won't forever tolerate companies that put them out on the street. And a company that thinks enough of its employees to provide them with health promotion and fitness programs won't forever tolerate bosses whose outdated attitudes sabotage the programs. So that kind of pressure continues to give way to a more supportive work environment. Meanwhile, the alcoholic worker has a resource near at hand to help him cope.

Stress

Everybody has read or heard enough about stress to understand that it is here all the time and too much is harmful and we ought to learn how to handle it. For most of us, what this means is that it's impossible to live without being exposed to a multitude of experiences that more or less constantly annoy or upset us, and if we can't cool it we're going to end up some day with a heart attack.

That's accurate enough as far as it goes, but it's really a lot more complicated than that. In the medical sense, and that's what we're talking about here, stress is a physiologic response to *all* experience; it goes on all the time at one level or another, and it is intensified not only by experiences that arouse feelings of annoyance, upset, and fear but also by those that evoke satisfaction and joy. Thus physiologic stress is not the same thing as stress in the dictionary sense of tension or strain, although the latter may commonly cause intensification of the former. The stress producing factors (such as heat, cold, fear, joy) are called stressors, and the body responds to stressors automatically, according to their intensity, with a series of changes in body chemistry. Thus a fright or a surge of anger may stimulate a succession of glandular reactions resulting in sudden, sharp increases in pulse rate, blood pressure, and blood sugar, and these in turn initiate resisting or compensating responses. Continued exposure to the same stressor may cause the resistance to wear down or become exhausted, creating a need for rest—a few minutes of planned relaxation, a quiet weekend, or a month's vacation, according to the severity and duration of the exposure—to restore normal function. The stress process was described years ago by Dr. Hans Selye of the University of Montreal, who called it the "general adaptation syndrome."[7] In popular terminology it is known as the "fight or flight response" that has been observed in animals and is often described by athletes as getting "hyped up" in preparation for a competitive event.

Stress Without Distress

In a book called *Stress Without Distress*, Dr. Selye pointed out that in considering stress-producing activity, the initial response is similar whether the situation is pleasant or unpleasant. "All that counts is the intensity of the demand for readjustment or adaptation," he said. "It is difficult to see how such essentially different things as cold, heat, sorrow, and joy could produce an identical reaction in the body; nevertheless, this is the case."[8] For a variety of reasons that Selye and other researchers have been studying for years, however, the adaptation or resistance process varies qualitatively with different stressors, and the same stimulus acts differently in different individuals. "Apparently the body can regulate its reactions through chemical messengers and nervous stimuli which either pacify or incite to fight," Selye explained, adding that some stimuli "act as tissue tranquilizers, creating a state of passive tolerance or peaceful co-existence with aggressors," while others cause changes mainly through the production of destructive chemicals. This is a greatly oversimplified description of an enormously complex process that physiologists have had a hard time understanding, let alone explaining. "The 'nature knows best' theory seems to be particularly obvious in the case of adaptation reaction," Selye concluded, with what may be as good a way as any of explaining the curious distinction between the kind of stress that is necessary to sustain life and the kind that tends to destroy—or the difference between stress and distress.

Because most of us spend our weekends and vacations substituting one set of stressors for another instead of avoiding exhausting excesses, we may be building harmful stresses and depleting responsive adaptation energies. Over time, this may increase the disposition to hypertension and added risk of atherosclerosis, heart attack, stroke, ulcers, and kidney disease. And because the work place in modern society is likely to be a hotbed of stressors, management of stress is a major goal of company health education programs.

First, workers are informed about the physiology of stress, so they will understand that the process goes on all the time, whether they are aware of the changes or not. They learn how prolonged or repeated exposure to stressors that produce intense responses can wear down resistance over time and result in illness. In the examination and discussion of stressors, they are encouraged to make lists of the events and circumstances in their lives, at work and at home, that are likely to cause intense stress response; in group sessions they may be encouraged to discuss these with one another, if they wish to. For many workers, the lists and conversations will suggest ways of avoiding or modifying the occasions that evoke stressors—setting the alarm clock 15 minutes ahead to ease the mad dash for bus or train, for example, or simply not responding to a challenging or argumentative assertion by an associate or family member when experience suggests that the outcome is likely to be a shouting match.

As in the case of alcoholism, many companies that have not yet seen fit to undertake comprehensive health promotion and physical fitness programs have nevertheless recognized the widespread incidence of stress among members of the work force and have hired or made available psychologists or counselors to whom employees may be referred, or may consult voluntarily, when symptoms of undue stress are observed or felt. Common signs include such things as unexplained and more or less continuous fatigue, joylessness or actual depression, withdrawal from contact or communication with others, frequent unexplained absences, deterioration in quality of work, irritability, listlessness—all or any of these behaviors may be signals that the worker is affected by stress. When these symptoms grow progressively more severe, the condition is often described as "burnout," a term that has been used increasingly in the last year or two to identify a condition psychiatrists refer to sometimes simply as "chronic stress."

In severe cases of stress or burnout, if conditions at work appear to be basic causes, a counselor may consider the advisability of recommending a change of assignment or work location for the worker. Where family problems seem paramount, a cooling off period, such as separate vacations or even a trial separation, may be needed to provide an opportunity for self-assessment in a relatively stress-free environment. Often the worker suffering from stress or burnout that has persisted for months may become despondent and down on himself, and the self-assessment process will then require a series of supporting sessions with the consultant to help restore self-esteem. In any case, systematic evaluation of oneself and all elements of the environment is the first step in stress management. Strategies for recovery include learning relaxation techniques and assistance in communicating with family members, associates, bosses, and others. In the company program setting, stress often can be limited or relieved by exercise regimens, diet, and education about health generally. As always, peer support when workers are exposed to the same or similar stresses is a priceless ingredient of the fitness formula.

Mental Illness

In a few cases when stress or burnout is severe and remains unrelieved for extended periods, susceptible workers may become anxious or depressed to a point where they can no longer function effectively. Frequent unexplained absences from work, forgetfulness, excessive anxiety or irritablity, and occasions when workers appear to be so withdrawn as to be unaware of their surroundings are signs that they may be headed for what used to be called "nervous breakdown." This was a layman's euphemism for mental illness or psychoneurosis, which causes normal thought and feeling processes to become disorganized and normal behavior to be impaired. Obviously, there is no clear line of demarcation between the temporary upset resulting from the kind of stresses we are all exposed to and soon recover from, and the behavior that suggests a significant change in the usual patterns of thinking and feeling. Similarly, there is no clear distinction, at the other extreme, between neuroses and the psychoses in which sufferers lose their ability to interpret reality.

In the company without either a medical department or a health promotion program that provides health education for everybody in the work force and creates general awareness of common health problems, the behavior of a worker suffering from one or another form of mental illness may go unnoticed or be passed over as relatively harmless eccentricity—until the loss of productivity becomes obtrusive. Then the person's boss or supervisor may start issuing reprimands or warnings—a course of action nicely calculated to worsen the condition, until the worker either quits or is fired, a result that means a loss to the company and a disaster for the employee.

Helping the Troubled Worker

At the health-informed and health-conscious company, in contrast, such behavior is more likely to be seen not as eccentricity or intransigence but as the treatable illness which it is. Within a few weeks of the onset of symptoms, in most cases, the worker is seen to be deeply troubled and is referred to a counselor. This may be a company psychologist or consultant, if there is one, or it may be the worker's own physician or clergyman. The losses and disasters may never be completely eliminated, perhaps, but with modern health education, bosses and workers will understand that mental illness is treatable, not blamable. They will know when to seek help and when to offer it, and they will know what help is available.

Because many of these troubled souls do go untreated and uncounted, and many more seek relief by joining self-help groups, reading popular do-it-yourself psychology books, and buying advertised over-the-counter drugs, there is no way to know the number who may need professional counseling or care. But some estimates may be made from a few facts that are known. The use of anti-anxiety prescription drugs, those that are ordered by physicians for patients who have consulted them for these and similar disorders, for example, has increased steadily from their introduction just a few years ago to a volume of more than 100 million tranquilizer prescriptions a year. And the annual total of prescriptions for psychoactive or "mind-changing" drugs is nearly 300 million a year—25 percent of all the prescription drugs of all kinds sold in the United States. "In any given year, 22 percent of the adult population of this country use some prescriptive psychoactive drug, 9 percent use at least some over-the-counter psychoactive drug, and 14 percent admittedly increase their alcohol consumption as a means of coping with distress," said Gerald L. Klerman, M.D., Harvard Medical School professor of psychiatry who served as director of the Alcohol, Drug Abuse, and Mental Health Administration in the U.S. Department of Health and Human Services during the Carter administration. "Most of these prescriptions are written by general practitioners and internists, not by psychiatrists, yet they are mostly prescribed for psychological problems. And there is little doubt that their use will increase for the foreseeable future,"[9] Dr. Klerman added.

With a quarter of the adult population known to be seeking help for some kind of mental or emotional upset, there obviously has to be a place in the company health promotion program to identify these people and start them on their way to the kind of assistance they need. In cases where the disturbance is not severe, company resources alone may be sufficient. Especially in larger companies, there may be room and time and staff for classes in stress management, relaxation, and such popular new psychological and communication techniques as assertiveness training and transactional analysis. In smaller companies, counseling may be provided or recommended. In any case, participation in exercise, fitness, and recreational activities may be a helpful adjunct to psychological treatment, and in every case the companywide health education that is essential to company health promotion will tend to create the kind of supportive peer associations troubled people are usually most in need of. But probably the best way to find out how these and other common health and fitness problems should be handled in company programs is to learn how some companies are handling them.

Chapter Five
MAKING IT WORK IN A COMPANY

There isn't anything unusual in the educational content of Control Data Corporation's health promotion program. It's based on well understood principles of health and human behavior—specifically, the principles around which Breslow and Belloc developed their rules for healthy behavior (see page 12). But educational content is the beginning, not the end. "The thing that *is* unique is tying it all together in the way we can do, in a consistent manner on a broad scale," Jack Heule, Control Data vice president for wellness programs, explained. "It's building a framework around the whole process and making it work in a company, making it effective for all employees, having a health promotion scheme that an individual will participate in and like. It's not just how you deliver an education program, but how you take it to a remote location and get top management totally involved, then sell it to middle management, play the role model, communicate, and get things done. We've had to tap a lot of kinds of expertise to make it play."[1]

The expertise was there to be tapped, or recruited, when it was needed, but first there had to be a concept. Control Data Corporation is a Minneapolis-based, worldwide technology and financial services company that pioneered and developed computer-assisted education systems in the 1960s. More recently, the corporation's capabilities were directed toward what it calls "self-health management" for two sets of reasons—philosophic and practical. The philosophic base is the deeply felt belief of the founder and chairman, William C. Norris, that the business of free enterprise is attending to the needs of the society that allows it to exist, with an ultimate result that can provide both profit for industry and a better quality of life for all of us. Based on that principle, corporate strategy has led Control Data to establish innovative programs in urban revitalization, small business support, education, employment, agriculture, energy alternatives, and, now, health.

The new health program matches the philosophic with the practical. Control Data's computer technology could readily be directed to analysis, information, and evaluation functions related to health. But there was also a more urgent need to become involved in health. Like other corporations, Control Data was paying huge sums for employees' health insurance; the cost had been increasing with the rising cost and complexity of health services throughout the 1960s and at an accelerating rate in the 1970s. "We wanted to see what were the principal causes of claims expense, and see if we couldn't educate people in some way to reduce part of that expense," Carolyn J. Kuhn, one of the developers of the health programs, told a visitor not long ago. "We wanted to work together with our insurers, and we asked them to give us an analysis of our claims. They came back some months later and said, 'Half of your people are suffering from this dread disease called "all other."' The claims that could be identified were about what you'd expect. So we went to the American Medical Association, the American Society of Internal Medicine, the American Academy of Family Physicians, and others, and their representatives worked with us to structure an appropriate program.

"Along the way, we were referred to the studies in prevention by Breslow and Belloc; these seemed to suggest a basic set of causes, so if we structured an education process around them, with an innovative look at what we might build as a solution together with medicine and other sources of expertise, it would give us a reach for the United States population. So what we're doing is legitimate for us as an employer and is also accurate and reliable, and it forms an effective partnership with medicine and other specialists in the private sector. The program marries professional expertise with the computer to manipulate large masses of data for the benefit of consumers, practitioners, and Control Data Corporation," Kuhn explained.

Actually, it was a lot more complicated than that. The process described here by Kuhn and others who played leading roles in designing and creating Control Data's self-health management program, now called StayWell, began in 1977. The first enrollments—groups of Control Data employees in New York and San Diego—came two years later. According to Kuhn: "We had an educational company, so we had design and publishing expertise here, but the content was purchased. We used external authors, external reviewers—subject matter experts. And what we tried to do was always stay within the margin of known medical fact—that indeed cardiovascular illness is related to these lifestyle risk habits, for example. We worked with 13 or 14 different medical specialties represented on our professional advisory board to determine what areas we should go into, as well as to reach toward the points in the Breslow study. The other components, like the voluntary participation we call 'cultural change support groups,' we worked out with another external consultant who had experience in another kind of lifestyle habit behavior change area. And we went ahead and designed that into our program. The other thing we've done that makes it unique is to build in an evaluation program. Of course, it's not like a double-blind clinical trial; in a program like this you can't deny access to benefits for anyone. But you can't offer it to all your people in different locations at the same time, so if you're prepared to start collecting data right away, you've got some leeway there to compare the experiences of participating and nonparticipating groups. For example, we have paired sites demographically by a number of different factors and can track data on several thousand employees, some of whom will not be enrolled in the self-health management program until 1983 or 1984."

The Facts Are Well Known

Control Data executives like to say that the education content of StayWell isn't all that unusual, meaning that the facts about health and health behavior are well known and widely accepted. But there is a logic to the StayWell content that isn't found in everybody's health promotion program, an orderly movement beginning with the presentation of information aimed at getting the participants to *believe* that their behavior can have an impact on their health—"to undo some of the things my mother taught me," as one of them said. The idea is that this knowledge builds readiness to deal with what follows, the content designed to teach *competence* to practice specific wellness behavior, and, finally, *compliance* with the desired behavior until it becomes a lifelong habit.

To achieve these goals, the program moves through a progressive sequence of education, motivation, and support activities—from an orientation session that everybody attends to a screening examination and health risk assessment for those who make the commitment to participate, then an interpretation of the resulting individualized "health risk profile" that enables the participant to understand his own health status and judge which ones of the education courses and organized activities he should—and wants to—take part in.

"What Control Data brought to this party was not a lot of health expertise as such," Jack Heule said. "It's our ability to take a complex task and systematize it, figure out what the elements are and go find the right resources to do what needs to be done. We're pretty good at that."

They must be. From the beginning, participation has been at an extraordinarily high level, with 90 percent of employees taking part in the orientation program and high percentages of most groups moving on to the screening and health risk assessment, and then continuing and attending education courses and group activities. Participation has remained at or near these levels among employees at the 12 Control Data work sites that have been added since the initial New York and San Diego groups were enrolled back in 1979. Skeptics have suggested that this level of performance may owe something, at least, to the fact that the corporation's interest and investment in health care and wellness are well known to employees, who could be expected to consider that failure to sign up might be regarded as disaffection. This could be true in some cases, executives are willing to acknowledge; the Control Data culture has certainly been oriented to health for the past several years. But in any event they are going to find out quickly whether or not this has been an important influence on levels of participation. After five years of design, development and testing, StayWell is now being offered for sale through Control Data's Life Extension Institute, an organization that has been offering medical and preventive services to industry for years.

StayWell isn't inexpensive. Depending on the number and location of employees and the nature of employment, the employer may pay from $400 to $700 per employee for the three-year program, including management for the entire period by Control Data's trained staff members. "StayWell is a three-year effort on our part," said Lee F. Kremer, Control Data vice president for health care services. "We see that as enough time to create the knowledge and motivation required for the program to carry itself. You've got to keep working at it until you change that culture. We're going into these work sites with the goal of changing individual lifestyles, and changing the work site culture—something that may include junk foods in the cafeteria, smoking, working long hours under stressful conditions. We want to teach people how to change that culture — how to cope with conditions that can't be changed and change the ones that can be. You can't do that quickly. It's a long process involving continued education, positive reinforcement, and peer support. It requires time, but we'll end up putting ourselves out of business if we're successful, because the people will take it away from us and carry on the program themselves," Kremer said.

Health Risk Profile

While it may be true that the high participation rate in StayWell has been possible because all the elements are carefully integrated and packaged in a fashion that can be transplanted and, as Jack Heule said, "made to work in a company," if a single component had to be selected as the chief reason StayWell works, it probably would be the Health Risk Profile, which combines the focused effectiveness of Dr. Lewis Robbins' person-to-person Health Hazard Appraisal (see page 13) as a means of motivating people to change their health habits, and the Control Data computer technology that makes it possible, in effect, to "mass produce" valid individualized health risk assessments for groups that may number several thousand employees at a single work site. The assessment consists of height, weight, and blood pressure measurements, blood chemistry examination, and an extensive personal, family, and health history and habit questionnaire that is completed by the employee, then combined with the measurements to produce the computerized individual risk appraisal. In workshop sessions, employees are then instructed in the interpretation of their profiles, which describe health risks based on individual findings and histories and indicate the ways risks can be reduced by lifestyle changes.

"The Health Risk Profile is designed to provide you with information on the general statistical probability that particular serious health problems may reduce the length of your life," says the explanation employees receive with their health risk printouts. "While this information is individualized because it is based on your answers to the questions in the Health Risk Profile, it is only a statistical approximation of the specific risks. It is not a substitute for the examination and testing done by a health professional. The computer printout is relatively simple to read and understand. Its purpose is to estimate your theoretical risk of dying from any of the 12 leading causes of death for people in your age, sex, and race group over the next ten years. But the most important point to be considered is the positive aspect, that by changing certain behaviors you may be able to extend your life expectancy. This Health Risk Profile and personalized wellness prescription are directed toward your taking personal action in your self-managed program toward attainment of a higher state of health."

Employee literature and company executives emphasize the complete confidentiality of all the information included in the profile. "The employee's right to privacy is maintained in all phases of the program, and each individual decides whether to share the profile results with a personal physician," says the instruction to employees. The same thing is true of a counseling and referral program for CDC employees that was initiated several years ago and is separate from StayWell. Described as a telephone "hot-line" service, this Employee Assistance Program (EAP) is staffed by professionals with backgrounds in clinical psychology, social work, counseling, and industrial relations, as well as medicine. Counselors staff a 24-hour a day service handling appeals for any employees or family members needing help with problems of drug or alcohol abuse, finances, legal entanglements, marital or family difficulties. The employee or family member needing help may make an appointment for face-to-face counseling, if this is needed, and many of the problems are handled by referral to community resources.

A supervisor who believes that personal problems may be interfering with job performance may suggest or encourage an employee to seek help, but it is left entirely up to the individual to decide whether or not to use the service, and in any case contact with the employee assistance service, as in StayWell, is completely confidential. "We have the privacy principle so drilled into our employee programs and our managers and employees that confidentiality is simply not a problem here," Jack Heule said. "The use of EAP is impressive, and it proves that employees believe in its confidentiality. I have never heard of any kind of violation of it."

Like StayWell, the Employee Assistance Program is marketed by the Life Extension Institute; the cost is $20 a year per employee. "Periodic reports are given to the employer on the *numbers* of employees using the program," a Life Extension Institute release says. "No names or other identifying information are included in the employer's report. If an employee wants the employer notified of his or her progress in resolving problems, or if special considerations that affect work need to be made, such as a leave of absence, arrangements are worked out by the employee."

A Powerful Motivator

Counseling and referral for treatment through the Employee Assistance Program is the way most cases of alcoholism, drug abuse, and mental or emotional disturbance are handled for Control Data employees. Except as these problems may be highlighted in Health Risk Profiles and thus alert employees to the need for help and motivate them to seek it, StayWell addresses these behavior problems in education courses rather than by individual or group counseling. But the Health Risk Profile can be considered a powerful motivator. Thus the profile of an employee who reported taking an average of 25 to 40 alcoholic drinks a week and smoking a pack of cigarettes a day would be informed in his or her profile printout that drinking has increased the risk of dying of cirrhosis of the liver 2½ times compared to the rest of the population, and smoking has doubled the risk of dying of atherosclerotic heart disease. Both risks can be reduced to the normal rate for the population by abstaining, the profile would point out.

Under the circumstances, it seems likely that the 8 to 10 percent of Control Data employees the National Institute on Alcohol Abuse and Alcoholism might have identified as heavy drinkers (see pages 45-46) may have put the brakes on their progression to alcoholism in the two years of StayWell. And it is obvious to a visitor at Control Data headquarters in Minneapolis that the company culture, as they like to call it, does not include a visible concentration of smokers. "Smokers around here are kind of conspicuous," one employee said. Ashtrays, if there are any, are out of sight.

Also not visible on the premises is any evidence of the corporation's dedication to physical fitness—outside the cafeteria, that is, where fresh fruits, vegetables, and salads are featured and hamburgers are frowned on. But the visitor interested in StayWell is not taken at once to a gymnasium or exercise room where perspiring middle-aged men and women are clocking up mileage on treadmills or stationary bicycles or toiling at muscle-building machines. He is not shown a running track on the roof or around the parking lot. Instead, he rides in an empty elevator to the eleventh floor, where he is told that "half the people on this floor walk up eleven flights of stairs once or twice a day, and people look sideways at those who take an elevator up one or two floors."

It isn't that StayWell doesn't comprehend fitness, or exercise, but the concept is to make healthy exercise a part of normal living, not an exhibition or a chore. "Many of the voluntary activities are fitness related, and the education program includes a fitness life-style change course," Jack Heule explained. "We help you design your own personal fitness program and ongoing support-group demand to help you maintain it. We emphasize the concept to a degree that encourages you to make a personal commitment, and then we'll help you find ways to go and do it. It would help if we had showers here, but we can't view that as essential. In-house facilities are more inclined to get those who are already hooked on fitness anyway. We want to reach the other 80 percent."

The fitness-related voluntary groups generate the peer support that is essential to maintain interest and participation for the long term. "The test of a fitness program is where you are three years—maybe five years, or ten years — later, not at the end of a class," Heule said. The activity groups don't just happen, and they aren't laid on by the corporation, but they may owe something to both spontaneous combustion and corporate encouragement. The process was described by Royce Pipes, who as corporation director of health programs has responsibility for recruiting and supervising Control Data's project managers in the field. Project managers install and direct StayWell at the work sites, and their jobs usually turn out to be part administrator, part health professional, part sales manager.

Pipes, who was trained as an engineer and is more at home in front of a chalkboard than he is behind a desk, charted the growth of a StayWell project in two curves. The curves represent levels of awareness and interest in StayWell, he explained. The first curve, A, rose steeply at the left side of the chart, then flattened out and began to decline as the line moved toward the right side. Curve B started below A at the left and rose gradually but steadily up and across — presumably to vanish at the upper right corner of the chart. "The curves show two kinds of activity level, two effects," Pipes said. "The program begins with a very intensive awareness effort. It's a new idea. Lots of enthusiasm at the front end. Orientation, screening, Health Risk Profiles—all useful by themselves, and also effective awareness instruments. That phenomenon, shown in curve A, declines with time—probably in the first year—and begins to fall off. The B curve represents awareness at the grass roots level. The more you can build this up, the higher the interest and activity will be. But A is my StayWell program, and B is the participants' ownership. The upper curve is my people driving this enterprise: 'You ought to know about this. You ought to do this. You ought to come to this class.' The other curve is their own self-generated level of interest, led by the informal opinion leaders in the work place, fueled by peer pressure."

Group Activities

The volunteer activity groups may come together and stay together for any purpose, as long as it's consistent with the health goals of StayWell, Pipes explained. Thus there are groups of dieters, and there are also groups experimenting in low-calorie cooking. There are groups of runners or joggers, and there are also groups that take walks on their lunch hours. There are groups that plan and take canoe trips, and groups that meet to help one another quit smoking, groups that lobby for bike racks so they can ride to work and groups that want space in the parking area for volleyball. The number and nature of the groups depends on the number of creative volunteer leaders who can be identified and encouraged to take the initiative. Among the 2,500 employees at Control Data's Arden Hills work site in the Minneapolis area there are 50 active volunteer groups meeting before and after work, during lunch hours, and on evenings or weekends, Pipes reported.

"Project managers are always looking aggressively for group leaders," he said. "You won't grow very many of your own, but they're there. The opinion leaders were there before you got there, and you can find them. You go out and talk to employees, and you ask them, for example, 'Who typically around here is always first to pick up a new fashion? Who organizes the summer picnic, or the winter dance, or the sports club, or whatever?' And because at some point you're going to start soliciting volunteers, you ask initially, too, 'Who should I stay away from?' Because you know that some of those who very likely are going to be among your early volunteers are the ones you want to avoid, those who for one reason or another would be a problem; they might want to lead, but nobody would follow.

"Throughout the program, we solicit volunteers for various kinds of things. We start by trying to generate task forces: 'Do you know anybody who's interested in walking at lunch?' You find some people who are interested in that, and you get them together and say, 'Why don't you organize a walk-at-lunch group, and go out among your friends here and get some people to join you?' The list of things to do is infinite. One of the first things a project manager starts doing is to make a needs or an interest assessment. What would people like to do? We're trying to be catalysts. Most of the ideas are not new, and they're not obvious, either. For example, we have a square-dancing group—in New York City. If I had to train somebody to go to New York, you can bet that organizing square dancing would not be the kind of talent I'd look for. But if it's in concert with StayWell, if it's something they want to do, they're more likely to stay with it, and it sounds like a good idea to me. We recognize that you've got to have a fairly high fun quotient in something like StayWell in order to get people to stay with it." Health promotion professionals don't like to emphasize the point, but the fact is that some motivation depends on scaring people — talking about dying because of what they do, and the number of years their behavior takes off their lives. But you can only scare people so much, or so long, and it isn't long enough. The negative emotions have a place, but the positive emotions last longer, Pipes believes.

The Basic Skill Is Management

So Pipes' administrator-health professional-sales manager types have to be part Cassandra and part Pollyanna as well, it would appear, and that isn't too far off the mark. Actually, they haven't been as hard to find as one might expect. Pipes has recruited talent from other fitness and health programs, and from organizations such as the American Cancer Society, the heart association, lung association, and so on. "They all have a strong interest in health," he said. "Most of them have some academic background in psychology, social science, counseling—something doing things with people." But the basic skill requirement is managerial, Jack Heule suggested. "We overkilled with talent at first," he said. "The experience was all in academic and public health fields, and this was their first exposure to the corporate plant environment, where they had to deal with the power structure, manage their time, find out who the key people were and get acquainted with them. But now we have the training of these people pretty well organized. You write it all down, get it out there, and tell them precisely what to do—and then they'll come back and tell you what's wrong with it. And that's a continuing process."

One of the things they said was wrong to begin with was the idea that executives and rank and file employees could be enrolled in the same programs, go to the same classes, and participate in the same groups together. That may be an acceptable style for businesses small enough to maintain a true family atmosphere, but it doesn't work in corporate settings, where the development and preservation of executive hierarchies is inevitable and probably necessary to the processes of orderly management decisionmaking. "You can't get a vice president to go sit in a stress management class with his employees, or even go to a weight reduction or smoking cessation activity with his peers," Jack Heule had found out. "So to get executives involved takes something different, and it is very important to have them involved so that you do get the kind of support you need from them at a work site where you're competing with a lot of other programs. We have to fight for conference room space, along with everybody else. We're after management's attention to support our plan to modify the menu in the cafeteria."

But there's more to the logic of a separate program for executives than protection for hierarchical separateness and approval of menu changes. "People who are highly educated, self-starting, used to setting plans and objectives and going after them, and monitoring their own performance, can usually do pretty well in a self-instruction program; just lead them to it, and they'll do it," said Murray Naditch, Ph.D., a clinical psychologist who is director of design and correlation for StayWell and has been chiefly concerned with designing the program's system for evaluating results. "Other people may require an instructor to lay things out for them and provide assistance and encouragement," he added.

So the StayWell Executive Program begins with the same Health Risk Profile that is given to all employees, but the interpretation is presented in a person-to-person consultation rather than a group session, and the education courses are not standard classroom fare but a series of stylishly written and handsomely illustrated booklets for executives to read and consider on their own time. The executive also makes his or her own "habit assessment" in diary form and, with a consultant, examines how present habits and lifestyle practices may influence the risks identified in the profile. Then the executive and consultant together establish goals and plan a course of action to produce desired changes.

Getting Families Into the Act

Another reason for the separate executive program is the nature and extent of participation by spouses. Getting families involved is important for all employees. "Initially we thought we could get families in by having the programs open to spouses," Naditch explained. "The initial spouse participation wasn't as strong as we wanted it to be, so what we've done is to focus our efforts to do spouse promotion in a couple of sites, and that's worked pretty well. We scheduled more convenient screenings, made things more accessible, developed materials for employees to take home, made it easier for spouses to come in and be part of the program. In some cases we've been able to increase spouse participation significantly," Naditch added.

But overall spouse participation at about 20 percent wasn't good enough for the kind of home support that was seen as essential for success of the executive program. "We could foresee a familiar kind of home-work dichotomy developing," Carolyn Kuhn related. "So we designed a special kind of promotion: gourmet dinner for executives and spouses together. The goal is to prolong life, protect the family, teach you how to manage stress together. The program has worked extremely well so far. It's often been the first opportunity for the spouses to share in a company activity. They're partners in it, and they're absolutely excited about it. It's one means of coping with the tendency of our lives to become compartmentalized, with resulting fragmentation of families and relationships and divided loyalties."

Integrating work place, social, and family influence is essential to sustained success of the program, Control Data planners believe. "I know from having done clinical interventions in this field for some 15 years that you can pull people out of their environment and make changes in their behavior," Dr. Naditch said. "You can work with them for six months or a year and accomplish major changes. Then you put them back in their environment, and pretty soon they're back where they started, because all the norms of behavior in the environment push people in an unhealthy direction. But by modifying the *environment* you can increase the probability of being able to maintain that behavior, because it becomes positively normative. You get a social movement started. People don't necessarily get into these programs because they want to improve their cardiovascular fitness. Most young people think they're never going to have to do that; they think they're never going to get old. But they may join because they want to play volleyball and meet some secretary who's attractive and who they know is on the StayWell volleyball team. And then everybody on the volleyball team is talking about how far they run, or swim, or whatever. So the first thing you know they're buying running shoes and taking up running. Or they may hear talk about nutritional cooking and get involved in that. People may join these groups for the wrong reasons, but eventually it becomes normative and prestigious and socially integrative, and it becomes literally a social movement."

Creating a social movement and changing a company's culture aren't goals that can be accomplished in a few months or a year. That's why StayWell is a three-year program. And over a five-year span, Control Data executives are convinced, lasting changes in habits and behavior patterns can be brought about and will continue to produce a return on the company's investment in StayWell—not only in reduced health care and health insurance expense but also in diminished absenteeism, increased productivity, and improved employee attitudes toward their work, toward the company, and toward themselves and their lives.

All these results are measurable in the evaluation system that has been designed into the program. "What we're looking at is to what extent the program is working for which kinds of people under what circumstances and to what effect," Murray Naditch said. "As we move into the future, we foresee the possibility of developing a computer-managed system that looks at individual differences, matches program interventions with those differences, and tracks people's changes in behavior as they move through the program. We are looking at changes in health attitudes, behaviors, and psycho-physiological indices, as well as measures of health care costs, absenteeism, and productivity. And we are using a variety of quasi-experimental designs, so that any inherent weakness in one method can be handled by others. The evaluation will also aim at finding out how people perceive the corporate support for health promotion and examining their attitudes toward the program and their beliefs about health," Naditch concluded.

The chief instruments for providing the computer with data on employee behavior and attitudes are personal lifestyle diary and health attitude survey forms completed by large samples of participating and nonparticipating groups. These are repeated over time so changes can be tracked. "We have done something unique with the diaries and health attitude surveys," Carolyn Kuhn reported. "That is, we have comparative samples of the U.S. population taken for us—using our own diary and survey forms—by Arbitron, the national opinion survey consultants, so we are not just measuring ourselves against ourselves. We were concerned that the Control Data culture might be different, and we were reassured to find that Control Data employees' attitudes and behaviors in groups measured before StayWell was introduced were identical with those of the population generally. We plan to repeat the national sample surveys periodically, so the differences that show up will be valid statistically."

About the only statistic that hasn't yet been entered into Control Data's computers is the number of U.S. companies that are ready to pay $400 a head and up for their employees to find out everything there is to know about their own health and do everything there is to do to make it as good as it can be and keep it that way. The statistic isn't going to show up tomorrow, but whenever it does and whatever it is, the way things are going it may be that $400 apiece for long-term change in employee's attitudes and practices would turn out to be the greatest bargain in health values since Lazarus came forth from the grave.

Chapter Six
WHO COULD DO IT BETTER?

At Baylor University Medical Center in Dallas, TX, there is a running track on the roof of one hospital building, where staff members and employees dedicated to fitness can run up their daily mileage quotas without risking pavements that threaten their arches or traffic that threatens their safety. There is an exercise room fully equipped with stationary bicycles, treadmills, barbells, and muscle machines, where fitness enthusiasts can work out as strenuously or as easily as their inclinations may dispose or the medical center's fitness director may recommend. In any case, there are shower and locker facilities permitting the most energetic of the exercisers to return to their desks or offices regroomed as well as relaxed. And the classrooms and lecture halls in the education complex at the medical center feature sessions on weight control, stress management, and smoking cessation, as well as clinics for physicians seeking new knowledge and surgeons learning new techniques.

At the Skokie Valley Community Hospital in a Chicago suburb there isn't a running track, or an exercise room, or a shower room, or a classroom. There is only a crowded little office—actually a patient's room with the bed moved out and a desk and chair moved in—headquarters for the hospital's Good Health

Program. Like the health education center at Baylor, however, Skokie Valley's Good Health Program is influencing the behavior of hospital employees and their families. Prowl the corridors after the busy routines of the hospital day are completed, and you may find an exercise group bending and stretching in a visitors' lounge, or an anti-smoking group watching a slide-film presentation or movie in the cafeteria.

And in many of the corporate headquarters office buildings and small businesses that ring the suburb, you'll find evidence of Skokie Valley's Good Health Program, which has reached beyond the hospital into employed groups in the area. Among the more prestigious groups that have been enrolled are the employees of Northwestern University's nearby Evanston, IL, campus and the university's professional schools in downtown Chicago several miles away—a fact that pleases Skokie Valley trustees and Good Health Program director Don Zeigler.

The Good Health Program didn't just happen. Don Zeigler told how it got started. "Jack Mabley and Jack Meyerhoff, two of our trustees, and Ralph Hutchins, administrator, were the people who wanted a pilot program," he said. "Our board represents the business community, and they are very concerned about health care costs. Jack Meyerhoff of the Brunswick Corporation was told by his board of directors to cut $1 million of the $15 million the corporation was spending on health care. How can you do that? There's only so much you can do with utilization review and claims management. You can't cut benefits to employees. The only thing left is lifestyle. This seemed to be a promising area, and Jack Mabley, a *Chicago Tribune* columnist and an idealist, had the idea that a hospital could have a key role in demonstrating that the private sector could have an impact on lifestyle. This was three or four years ago, at a time when there was a very real threat that a national health program might be enacted that would be very expensive and wouldn't really address the issues, especially not the area of lifestyle. The feeling was that the private sector could do it better than the government could. And who could do it better than a hospital, with its strategic location in the community, access to business, credibility, and the all-important professional expertise?"[1]

Doctors Didn't Come Quietly

The Skokie Valley board of trustees supported Mabley's idea enthusiastically, but it wasn't all that easy. "It took us a year to get off the ground," Mabley wrote later. "The first thing I learned was that physicians don't like the idea of others practicing medicine, especially when the others call it preventive medicine. The concept implies that physicians do not practice preventive medicine. I called a friend at [the] Blue Cross [Plan]. He was enthusiastic and offered to help us. I was vastly encouraged and asked him if he would come to a meeting of our medical staff and explain his ideas. In my naiveté I expected the medical staff to embrace this project. My friend from [the] Blue Cross [Plan] was introduced at the meeting, and you could cut the hostility with a knife. He is a pleasant and capable man, and he told what he had in mind, but rating the meeting on a scale of one to ten, with ten success and one catastrophe, I'd call it a one and a half."[2]

Eventually it all got sorted out. The trustees learned to talk about "fitness" and "lifestyle" instead of "preventive medicine," and the doctors found that less irritating. The hospital board budgeted $50,000 for the project and created a separate Good Health Foundation board to handle it. Jack Mabley got a grant from the McCormick Charitable Trust and Jack Meyerhoff got a grant from the Brunswick Foundation—the first of several grants in both cases. The president of a company that had done screening for industry came on the board, and, finally, Zeigler, who had worked with physicians and industry at the American Cancer Society, was recruited to put all the pieces together and get going.

Zeigler was familiar with the Health Hazard Appraisal and knew about the several computerized systems that have been developed from it, combining the results of screening, health histories, and lifestyle behavior questionnaires to provide motivation and direction for health education programs. Eventually, Skokie Valley chose the health risk questionnaire developed by Medical Datamation, one of many now on the market, and added a few more tests to the height-weight measurements, blood pressure, and cholesterol determinations that are standard on most of the forms. The education or health promotion content of the Good Health Program includes a no-nonsense stop-smoking sequence of six 1½-hour sessions beginning with a cold turkey, "quit right now" drill that is hard on the participants but has achieved a notable record of successes. Other classes are offered in nutrition, stress management, weight control, and exercise. The number and length of these sessions for employees to choose may be varied according to what the employer may want to offer, and pay for. One group, for example, the Evanston headquarters office of the National Assembly of Baha'is, the worldwide religious community, started with nutrition and moved along to stress management and weight control but skipped the stop-smoking sequence. Baha'is don't smoke. "At least," their realistic personnel director told a visitor, "not around here."

As Control Data did with StayWell, Skokie Valley started the Good Health Program at home. "Before taking the screening into industry, we decided on a trial run with hospital employees," Mabley explained. But that wasn't easy either. "How do you convey a program of preventive medicine to hospital people?" Zeigler wondered. "They are biased. They think they know it all. Everybody is a health expert, so we had some communication problems. We were all new employees, the young kids on the block, with no credibility in the hospital, coming in to tell the employees how bad their lifestyles were. They just didn't believe it—'Why do I need this?'—and they had some concerns about confidentiality." To help the program over these hurdles, Zeigler and his staff produced a slide-film featuring the unsightly and sometimes dangerous effects of bad eating habits, excessive smoking, stress at work, lack of exercise, and other common lifestyle trespasses, and telling how the Good Health Program can show the way to more enjoyable lifestyles—and longer lives. The presentation is more entertaining than hortatory, low key throughout, and takes only 10 minutes, and it resolved enough of the communication problems to enroll a respectable 46 percent of all hospital employees in the program. Because of the difficulties of hospital schedules, however, not that many completed the education courses—except for the 198 who signed up for the stop-smoking classes, 86 percent of whom went the full route.

The Trouble Is, It's Boring

Among hospital employees and in Good Health groups in industry, about 50 percent of female employees and 60 percent of males were still not smoking when follow-up surveys were made—self-reporting substantiated by testing—nine months to a year later. This is enough better than the national averages for smoking cessation groups (35 percent is par for the course) to suggest that Zeigler knew what he was doing when he insisted that his health promotion activity leaders had to be not only knowledgeable but entertaining. The trouble with so much health education material is that it is *boring*, and nothing could be worse than that, he said. "Our people have to be creative, talented entertainers, as well as experts in their fields. They use lots of displays—overhead projectors and transparencies, piles of transparencies—at every session to keep things lively and hold attention."

The stop-smoking sessions are a prime example. "You quit right now" is the opening shocker, and there isn't a dull moment from then on. The tricks include the usual displays of what smokers' lungs look like and the demonstrations of how quickly smoke permeates the system. "At the first class I said to myself, 'Hey, maybe this guy knows something,' " an employee whose wife had talked him into attending the clinic said of Joel Spitzer, the smoking program coordinator. "His enthusiasm made me want to quit to please him. I couldn't let him down. During the clinic I bitched and moaned, and Joel prodded and needled. He did everything but stand on his head. He got me through the first day, hour by hour, then the first week, and finally the first year."

Actually, the secret of the 50 and 60 percent stay-quits is neither Spitzer's black-lung demonstrations nor his headstands, but his use of employees like this one to attend the classes and provide the priceless ingredient that the Good Health Program, like Control Data's StayWell and most others, have found essential to success: peer support. "He gets former smokers to help," Zeigler explained. "They come in and tell how they did it and generate the attitude that 'I can do it too.' The first three or four days, the hard part, he sees the new quitters every day. He convinces them that they aren't just trying to quit. They already *have* quit, and he's just reinforcing the decision. Then at the end of the six sessions they have a graduation ceremony, and they all get diplomas. They make speeches about how they did it, and applaud one another, and get very excited. It sounds corny, but it works."

The smoking course isn't the only one that gets people excited. After seeing Zeigler's slide film and deciding that the Good Health Program was just what was needed for the Baha'i center, where use of health care benefits was excessive and turnover was a problem, Floyd Tucker, Ph.D., personnel officer at the center, had arranged for the program to be offered to Baha'i employees. He described what happened at the first nutrition class. "Everybody came," he said. "They had prepared special foods for us, like meatless chili and low-calorie pizza, so people could experience a different kind of food. That got them going, with people jumping out of their seats and asking questions."[3]

Tucker had heard about Skokie Valley's program when it was being introduced at Northwestern University and visited several classes when he was considering what might be done for the Baha'i staff, whose members come to Evanston from all over the world and often have a hard time adjusting to new ways of living. "As an educator, I saw immediately that they knew what they were doing," Tucker said of Zeigler and his people. "The instructors were well versed in their subjects, and up to date." Tucker and Zeigler discussed the special needs of the Baha'i employees and decided what components of the program would be most appropriate for the group of 150—representing a variety of cultural backgrounds, and having more women than men. "Zeigler visited us and familiarized himself with our practices," Tucker related. "Then he made a proposal which I presented to the National Spiritual Assembly, our governing authority. I told them what I wanted to do, and they approved the $5,000 expenditure I asked for." At first, they had planned to go right into the education classes, starting with nutrition, but after executives and department heads had seen the slide-film, they decided to offer the health hazard screening to employees who would pay for it themselves, at $44 per person. Many of them signed up.

They're Going To Change

The group moved on from nutrition to weight control and exercise classes, and then asked if they could have a special class in nutritional cooking. Within a few months, Tucker said, there was already evidence of what he considered to be significant behavior changes, especially in eating habits. "They come into the lunch area now with fresh fruits and raw vegetables," he said. "There was a lot of excitement at first, but what we're getting in the long term is that people may say, 'I still grab a doughnut once in a while, but every time I do it now, I think about it.' And if any program can do that, there's hope for it, because it means they are going to change. They're going to start cutting down on their use of salt, for example, cutting down on butter fats, hypersaturated fats. If we can do that, then we've accomplished something, and I really believe we have. I've had people say to me, 'I still eat my pizza, but I feel guilty about it.' In the long run they're going to change their eating habits, and that's what is exciting to me."

The prices of components of the Good Health Program vary with the number and length of the education sessions selected, Zeigler explained, but the basic prices, in addition to the $44 per person for the health risk index, interpretation, and counseling, are $750 for the six 1½-hour stop-smoking clinic sessions, including Spitzer's headstands; $880 for "Eating for Good Health," the six one-hour nutrition classes; $900 for 14 45-minute weight control sessions; $800 for five one-hour stress management classes; $180 for four one-hour "Introduction to Exercise" periods; $900 for 16 one-hour exercise classes, and $900 for 16 "Dance Fitness" classes featuring instruction in aerobic dancing. These prices are based on class sizes ranging from 10 to 30 persons in most cases—about what might be suitable for a group of 500 employees. To accommodate a group of 1,000, most of the courses would have to be offered twice, to keep class sizes manageable, and the cost would go up accordingly.

These prices turn out to be $55 per employee or family member, give or take—a bargain if it results in lifestyle habit changes that stay changed, or changed enough to keep some people out of doctors' offices and hospitals—a critical question the answer to which remains to be disclosed. Meanwhile, the Good Health Program recommends periodic revisits, to provide enrollment opportunities for those who were "previously unable to participate," as a Good Health protocol discreetly phrases it, and "to give previously screened individuals the opportunity to see improvements on tests as a result of lifestyle changes." The recommended revisit procedure includes repeating the health risk index questionnaire and screening, and repetition of the basic education activities or classes as indicated, as well as advanced or special interest health promotion activities if these are seen as needed or desired.

In any case, it seems clear that the Skokie Valley Community Hospital has already accomplished several of the goals it set out to accomplish when the proposals presented by Mabley and Meyerhoff were accepted by the board of trustees. For the hospital, these were: "to give balance to the hospital's goal to provide preventive, curative and rehabilitative services to the community; to develop and demonstrate a model community hospital primary prevention outreach program, with a positive impact on health hazardous lifestyles by teaching desirable steps to wellness, and to offer the medical community reasonable, effective risk reduction programs to which they may refer their patients." For individual participants, the goals were to help them "understand the major causes of those diseases related to lifestyle and which are susceptible to alteration by early detection and changes in personal habits," and to help them "become aware of those factors which may be hazardous to their good health and motivate them to adopt habit patterns which lead to a high level of wellness."

Among the goals that can't yet be considered accomplished were, for the hospital, "to provide a private sector demonstration project of cost-effective efforts to lessen costs, save lives, and reduce morbidity of the public." And, for employers, goals were to "provide a means to save lives and enhance the quality of life through improved living habits; to cut health care costs, reduce absenteeism, and increase individual well-being."[4]

Obviously, it may be years before ambitious goals like reduced morbidity and mortality can be credited to lifestyle changes brought about by health promotion programs, but it isn't too soon now to say that many employers consider that they are getting their money's worth from such programs in reduced absenteeism and improved morale and productivity. And it won't be long before these results can be measured in dollars as well as perceptions.

Whether that time comes early or late, however, it is an

astonishing fact of our time that business and industry, commonly thought to be hard-nosed about spending money, are accepting health promotion for employees as a matter of common sense and good faith and spending substantial sums without waiting around for explicit answers to the question of how much it costs and how much it saves. It was the businessmen on the board of trustees of the Skokie Valley Community Hospital who gave Mabley and Meyerhoff the money to get their project started. The program wasn't expected to make money, or even to break even, according to Zeigler. "The motive was altruistic," he said. "We wanted to develop a model and share it with the world, but to be very critical of what works and what doesn't." So far, it is working.

Chapter Seven
HOSPITALS AS CHANGE AGENTS

A company that lacks the resources or the willingness to invest in a full-scale, long-term health promotion program like Control Data Corporation's StayWell may nevertheless be convinced that a scaled-down effort to motivate employees to take better care of themselves, and help them learn how, would be worthwhile. For such a company there are many possible sources of service, information, and advice to be investigated. In addition to the many national organizations now offering health promotion packages that include screening, counseling, education, and fitness programs—many of which can be adapted to the employer's specific needs and capabilities (see page 127)—some of the services, at least, are often available in the company's own community.

A logical place to look first for help is at nearby hospitals. Not every place has a Skokie Valley Community Hospital, by any means, but there are hundreds of hospitals in the United States offering comparable comprehensive health promotion services to business and community groups, and there are many more with public education and counseling programs that might readily be available or adapted to meet the needs of employees. The thing for an employer or appropriate company representative to do is visit or call all the hospitals within a reasonable distance and find out what kinds of services may be usable. Chances are good that there will be a wide range of programs to consider. In a survey reported some months ago, the American Hospital Association found more than 1,200 hospitals that were providing programs for local businesses "designed to help keep employees healthy," the report said.[1] Since that time, many more hospitals have initiated programs, as the wellness movement has been gaining momentum.

The AHA has actively promoted development of these hospital programs. In a 1979 policy statement, the association declared that "the hospital has a responsibility to work with others in the community to assess the health status of the community; identify target areas and population groups for hospital-based and cooperative health promotion programs; ensure that persons who are apparently healthy have access to information about how to stay well; provide appropriate health education programs to aid those who choose to develop a more healthful lifestyle; and establish the hospital as an institution that is concerned about good health in addition to illness."[2]

A 1980 release from the AHA's Center for Health Promotion focused attention more specifically on business. "Businesses are undertaking extensive programs to help ensure the good health of their employees," according to the release. "Hospitals, too, are developing programs for their employees, and they also are adapting and expanding these services for use by local businesses. In the process, everybody emerges a winner: Businesses have the potential to improve the health and morale of employees; employees have an easy-to-access program enabling them to improve their health behaviors; and hospitals develop a new service which meets documented needs of businesses and their employees and which has the potential to generate new revenue for the institution."[3]

Wellness in Hospitals?

It figures. As Skokie Valley's Don Zeigler said, who can do the job better than hospitals, with their convenient locations, professional knowledge, and management resources? When the idea was first being proposed just a few years ago, however, before the interest in physical fitness had become so publicly visible, the educational concerns of doctors and hospitals were limited for the most part to teaching diabetic and heart disease and colostomy patients how to live with their disabilities. The notion that hospitals should devote any part of their limited resources to keeping people well, and thus emptying the beds they had to keep filled in order to earn the revenue they had to have to do all the things they were expected to do, was regarded as bizarre. Doctors, especially, were skeptical, to state the proposition conservatively. A young pediatrician who had been persuaded to spend part of his time as a school physician explained why. "It drives me wild to spend a morning at school looking down healthy throats and ears when my office is filling up with sick kids who need my help," he said. While cynics might say the reason for this is that the doctor gets paid for sick kids but not for healthy throats, it's actually a lot more complicated than that. The real reason for some doctors' disinterest in well people is deeply embedded in the medical *zeitgeist;* from his first day in medical school to his last day in practice, the doctor's attention and energies are focused on pathology. It is sickness, not health, that challenges his knowledge and skill, occupies his mind, and excites his emotions; except when it reappears following a treatment he has recommended and applied, healthy tissue is a bore. Well people are a diversion; sick people are the real world.

Given the disposition of the medical culture, it is remarkable that so many physicians have spent as much of their time as they have recently, helping to plan and staff the many health promotion programs that have been initiated in hospitals, businesses, communities, and in their own practices. This has been a response to the insistent demand for such services. The demand made itself felt in many ways; health food industries, nutrition classes, diet regimens, exercise groups, and stop-smoking routines multiplied all over the place; books and magazines on fitness and health crowded the bookstores and newsstands for space alongside those on sex and crime. For the first time since our grandmothers took gallons of cod liver oil, when vitamins were new and fashionable but didn't come in capsules, good health was news. Inevitably, hospitals were among those that first became aware of the fact and moved to meet the demand. Those hospitals the AHA counted in its health promotion survey had been calling themselves "centers of health activities in the community," as most hospitals had been doing for years. They had franchises to protect, so to speak, and they got busy. Some typical hospital initiatives:

- *For Your Good Health* is a quarterly publication in newsletter format published by Health Education Publications, Inc., of Chicago and distributed to hospitals for local circulation in their communities. The name of the sponsoring hospital appears on the cover; some hospitals add the name or names of doctors on their medical staffs to lend professional emphasis, and provision also may be made for other local copy on a back page.

  The content of a recent issue included a leading article on tumors—what they are and what to do if a tumor appears or is suspected, and how they are diagnosed and treated; an article on skin and skin care; a dental feature; a news report on the patient package insert that comes with medications; and a caution about the safe use of prescription drugs. Articles are expertly written and checked for accuracy by a panel of distinguished specialists.

  The minimum hospital order is 2,000 copies; some hospitals buy as many as 50,000 copies for distribution to patients, former patients, donors and prospective donors, businesses, schools, churches, community leaders, and groups of all kinds.

  In some cases, industries or business groups pay for the publication and have it circulated to employees' homes, and in some communities hospitals and industry may share sponsorship. Thus in Aberdeen, WA, for example, the notation on the cover of *For Your Good Health* reads "Issued by Grays Harbor Community Hospital and St. Joseph Hospital in cooperation with Rayonier ITT Grays Harbor Division." The total U. S. circulation of the publication is in excess of 2.5 million copies.

Focus on Positive Action

- An unusual community service area at a shopping mall in La Jolla, CA, includes a health education center sponsored and staffed by Scripps Memorial Hospital a mile away. Called The Well Being, the center offers group and individual counseling on problems related to health and lifestyle practices and maintains a library of health literature and audiovisual media on exercise, nutrition, stress management, and related health topics.

Surveys conducted at the shopping mall and evaluation reports completed at Well Being seminars have demonstrated that the center is meeting community expectations and needs, according to Christine Brown, R.N., M.S., Well Being director. "The programs are designed to focus on the positive actions a person can take to maintain good health," she said. "We look at a person's health-related strengths and weaknesses, then try to focus on making the strengths overcome the weaknesses. Rather than dictate the decisions about a person's lifestyle, we try to provide the opportunity for the person to participate with professionals in decisionmaking."[4]

Well Being activities have included a cardiorespiratory fitness program with public blood pressure and pulse readings, hikes to promote walking as a form of regular exercise, and a public forum on the benefits of exercise, in addition to the seminars.

Scripps Memorial has another program, called Prevention Plus, offering health evaluation and exercise, nutrition, stress management, and preventive counseling services.

- Salem Hospital, Salem, MA, several years ago started offering pre-employment physical examinations and routine physicals for employees of local businesses. The services were planned initially to help employers meet requirements of the Occupational Safety and Health Administration, and other services have been added since in response to employers' needs. Now companies in the area are contracting with the hospital for periodic industrial nurse visits to identify employees who need health counseling or referral for medical care. The nurse checks on employees' health following absence for illness, for example, and may also conduct group screenings for blood pressure, urinalyses, vision and hearing tests, and other determinations as indicated. Stress management counseling is offered, and assistance may also be provided for employees with mental health or personal problems. Such counseling is paid for by the employer on a fee-for-service basis, reported the hospital's preventive medicine department, which manages the program.

- In addition to 24-hour emergency care for work injuries, Family Hospital and Nursing Home of Milwaukee, WI, offers companies a range of health assessment services on a fee-for-service basis. Physicians in the hospital's occupational health department are assigned to companies, as far as possible matching the physicians' specialties and the companies' needs. The initiative for establishing the program came from small businesses in the area, a hospital report said, and from a follow-up poll of local business and meetings with an employers' association to determine specific needs and extent of interest. Twenty-five companies signed up for one or more of the offered services, it was reported.

The Health Hazard Appraisal

- Methodist Hospital of Indiana, at Indianapolis, where the Health Hazard Appraisal was invented by Drs. Lewis Robbins and Jack Hall, provides the risk screening service to industry in the hospital's area. Through its department of prospective medicine, the hospital organizes education and counseling services for employees based on the results of the appraisal. Interestingly, according to Scott Sheedy, director of the department, the Number One group at risk in many of the industries screened in the hospital program has turned out to be one that has nothing much to do with medicine and is not ordinarily subsumed among problems related to health. It is employees who drive to and from work every day without fastening their seat belts, as disclosed in the answers to one of the lifestyle questions in the Health Hazard Appraisal. "The loss of an employee who gets killed or injured in an accident is just as costly to the employer as the loss of one who has a heart attack," Sheedy said, and there are more employees risking accidents, so that in many cases the most important need is for safety instruction and motivation. Other major needs of Indiana employees have been for hypertension control and regular exercise regimens, Sheedy added.

Outside the hospital's service area, the department offers employers the Health Hazard Appraisal and interpretation, and employers contract with others for education, counseling, and other needed activities. In some areas, Sheedy has arranged with Blue Cross and Blue Shield Plans to offer the Health Hazard Appraisal to employed groups enrolled in the Plans.

As owner of the Health Hazard Appraisal, which is a legally protected system, Indiana Methodist Hospital on occasion has licensed others to use it. One such licensed user is the Life Extension Institute, now a part of Control Data Corporation. Another is the Canadian government, which is authorized to use the system in Canada under conditions stipulated in the licensure agreement. Because the system has been widely publicized, however, it has also been appropriated and adapted by a number of unlicensed users. This is distressing to Indiana Methodist not only because such use threatens the revenue to be earned from the substantial investment involved in the system's development, Sheedy explained, but also as a matter of principle, and because unauthorized users may be advising groups on the basis of data that are no longer valid. "We keep the system current by updating the data base periodically," Sheedy said. "Groups using a computer program based on outdated statistics may be getting bad advice."[5]

- Like many others, the health promotion program at Pacific Medical Center (PMC) in San Francisco began as an occupational health service offering pre-employment physical examinations and advice on accident prevention. But as an occupational health physician observed, "There is often a fine line between work-related problems and those which are not." PMC's department of occupational health now contracts with San Francisco companies for a wide range of activities, from stress management counseling to noise control, and including periodic physical examinations related to specific job requirements. "Routine physicals are not cost effective," said Dr. Linda Clever, director of the program. "There is no reason to do a hearing test on an accountant, but there certainly is a need to do one on an engineer."

Among the more unusual assignments the department has handled were a desensitization program for a group of bank employees who were on the job during a robbery and a cardiopulmonary resuscitation course for two truck drivers serving the Golden Gate Bridge, who were often the first ones to reach the scene of accidents. And a department store whose employees had a lot of back injuries asked PMC to train them in lifting heavy boxes. "We're here to provide whatever services are needed," Dr. Clever said.[6]

Wholesale Health Promotion Packages

The number of hospitals offering health promotion services to companies in their communities is increasing rapidly, not only because of growing interest in the services themselves but also because a number of enterprising organizations are offering complete packages of risk assessment, health education, and lifestyle change programs that hospitals can provide local businesses. The entrepreneurs furnish the organizing principle, the educational materials, and the behavior modification techniques, and train hospital personnel to carry out the necessary functions—making the hospitals, in effect, retailers of health promotion services for employers in their communities. "We save our hospitals the time and the substantial investment it takes for the hospital to develop its own health promotion program from a standing start," said Dick Peterson of Health Central, Inc., a Minneapolis-based hospital system.[7]

Three of the hospitals in the Health Central system are selling "High Level Health," the health promotion plan created by the Sun Valley Health Institute, whose staff of exercise consultants, nutritionists, physicians, and psychologists is well known to hundreds of U.S. corporation executives who have been through the organization's executive fitness program at Sun Valley, ID. "We believe that the community hospital is the single best source to guide people to achieving healthier lifestyles," said Gary Steinbach, president of the institute. "Hospitals provide the credibility, health professionals, diagnostic services and facilities necessary to advance the health promotion movement on a sound basis. Our goal in working with community hospitals is to help prevent the large capitalization costs of developing a program from scratch, along with inherent time delays associated with this process. We train people from the hospital to put on the program, utilizing the hospital's ancillary services and facilities for testing and education."[8]

The Institute has a formula for estimating unnecessary health care costs and projected savings within an organization, and the program emphasizes the importance of tracking program outcomes. "A marketing and sales program is designed for each hospital in order that the hospitals may make their health promotion efforts an ongoing and robust business activity," said Steinbach. "This marketing tool is available for hospital sales work with the local business community to demonstrate cost containment potential."

Speaking for the Health Central hospitals, Peterson said the cost containment approach is essential. "In times like these, business isn't accepting new programs on the basis of talk about improved employee morale," he remarked, noting that initial acceptance of the programs in some of the hospital communities had been slow. "We have to emphasize the cost containment aspect, as well as the fact that the hospital, with its professional and caring people, is the most appropriate resource for a program concerned with health activity."

A comprehensive wellness program some hospitals are offering to businesses and community groups is the one that was developed by Robert F. Allen, Ph.D., president of the Human Resources Institute of Morristown, NJ. Dr. Allen is a psychologist whose interest in wellness emerged from studies of cultural influences on behavior. In projects involving migrant workers in the South and delinquent young people in northern city ghettos, he developed what he called the normative systems change process. "Both these projects rose out of cultures that were mired in whole systems of expected behaviors that had kept people from achieving what they wanted," he explained. "They only had to discover their own power together to make change. Soon the system was being used in many situations, in groups ranging from small family units to large corporations, schools, and communities. The primary aim, no matter what the application, was to help people find a way to help themselves, to determine the course of their lives."[9]

Changing the Norms

The normative systems change process that helps people help themselves to better health behavior is Lifegain, and it begins by gathering baseline data from which later gains can be measured and by which the program can be tailored to specific needs. "The tailoring process is carried out by task forces organized around particular areas of concern, such as developing new facilities or organizing recreational programs, or dealing with smoking locations," Dr. Allen said.[10] Two of the hospital groups that have adopted Lifegain, first for their own employees and staff members and then for application elsewhere, are the Samaritan Hospital System based at Phoenix, AZ, and the Swedish Medical Center of Englewood, CO. Dr. Allen and Gary Adamson, director of the Swedish Wellness Center, together with Richard Bellingham, director of health promotion for Samaritan Health Service, described how the program is introduced to the entire work force in a series of workshops. "Through these high-involvement workshops, running from two hours to two days depending on the time and purposes, everyone has an opportunity to learn about the immense impact of health practices and health cultures on their lifestyles, to identify their own personal situations, and to decide on plans for change," they said. "The workshops usually start with the top of the hierarchy but eventually are offered to all employees."

Phase III, following the data gathering and workshop processes, is the time for initiating change, "putting into daily action the plans begun in the workshops," wrote Dr. Allen and his associates. "To support them in their change efforts, people are offered membership in support groups. Some are general support groups dealing with a variety of health concerns. Others are specialized groups with a focus on a particular problem. In addition, there are specialized intervention programs, self-help programs for individuals, and task force activities that touch a cross section of employees in a wide variety of programs. The task forces are most important, for they address themselves to organizational change problems."

Finally, Dr. Allen and his associates wrote, "Phase IV is a period of sustaining change, evaluating the program, renewing enthusiasm, and extending the program to other locations or to other problems." It will require at least two years to develop data about normative change such as weight loss, cardiovascular improvement, stress reduction, and other health areas Lifegain is focused on, Dr. Allen concluded. "It isn't easy, particularly since many of the approaches to change used in our culture in the past have actually inhibited and interfered with positive change processes. Campaign-type efforts, reliance on finger-pointing and blame-placing, use of scare tactics, and focus on single elements within the system all lead to failure rather than success, and to the feeling that there is not much that can be done about the problems that confront us. Our goal must be to look at our health culture in a broad perspective, and to work in a positive and constructive way to change those norms that affect us—a way that can give people confidence in their own ability to change things for themselves."

Chapter Eight
PREPAYMENT PLANS AND OTHER RESOURCES

Before a company decides on a source of supply for its employees' health promotion program, another possible source that should be investigated is the Blue Cross and Blue Shield Plan in the area. The Plans all have useful health information and educational materials that are often available either without charge or at minimum cost. Many of them offer screening services and planned educational programs based on company needs, similar to those marketed by others. In Ohio, for example, the several Blue Cross and Blue Shield Plans have joined together to offer employers a comprehensive program called Health Choice, which begins with an evaluation of the group's health needs and priorities and includes courses and educational materials focused on producing desirable lifestyle changes. Blue Cross and Blue Shield of Minnesota provides consultations to match health problems with health and fitness programs available in the community. Because of the importance of heart disease as a cause of illness and disability in the United States, many of the Plans offer blood pressure testing services, either at Plan offices or in vans that tour Plan communities; some Plans have computerized blood pressure machines that can be taken to work sites, where Plan health staff members conduct employee surveys. For those identified as being at high risk for heart disease, Plan education programs emphasize the importance of regular physician visits and continued compliance with recommended medication and lifestyle behavior change.

Blue Cross and Blue Shield of North Carolina pioneered several years ago when the Plan introduced a program called Education and Screening Examination (EASE), which was developed jointly by Dr. Siegfried Heyden, professor of community medicine at Duke University School of Medicine, and Dr. William DeMaria, Plan medical director. Conducted at the work place, the intensive EASE screen is aimed at identifying risks of hypertension, obesity, diabetes, smoking, and cancers of the cervix, breast, mouth, colon, bladder, and prostate. Also cooperating in the program's development were representatives of the management of one of the state's large employers, Cannon Mills, where 10,000 employees were screened in the first year of the program. In a report issued at the end of the year, the Plan said that 15 percent of the employees screened were found to have high blood pressure and were referred for medical care and follow-up treatment. Eight employees per 1,000 screened had previously undetected diabetes that could be controlled by weight reduction and dieting. Twenty cases of early cancer were discovered by the doctor-nurse teams conducting the screening, it was reported. These patients were treated at an estimated cost of $2,000 each, and the prognoses were judged to be favorable. Although the ultimate outcomes would not be known for a matter of years, Plan physicians estimated the cost of treating cancers undetected in the early stages would be at least $20,000 per case—an amount that suggests a measure of the cost-effectiveness of preventive screening.

In 1979, Blue Cross and Blue Shield of North Carolina invited 500 employers in the state to a conference on health promotion in industry. More than 200 companies sent representatives to the conference, where the reasons for having employee health education and fitness programs and details of operations, results, and cost were examined. In a recent survey, more than half of all the companies—including those that came to the conference and those that didn't—had initiated some of the health education or fitness activities that had been discussed, H. C. Cranford, Jr., vice president for public relations of Blue Cross and Blue Shield of

North Carolina, reported.[1] In 1981, Cranford said, the Plan and the North Carolina Governor's Council on Physical Fitness and Health would establish an annual Health and Fitness Award for business and industry in the state, recognizing outstanding achievement in promoting employee health. Among the Plan's many health education activities are a newsletter called Healthwise, a consumer's guide to medical care titled "Take Care of Yourself," and a new first aid manual. These and other health materials are distributed through literature racks located in doctors' offices, hospital lobbies, and work sites with 50 or more employees. More than 4,000 such racks installed throughout the state assure that information on good health habits is getting to a substantial segment of the population.

What Blue Cross and Blue Shield Plans Are Doing

To round out the picture of Blue Cross and Blue Shield Plan health promotion activities having particular interest for employed groups, the following paragraphs mention some of the things a few more of the Plans are doing. It should be emphasized, however, that the brief descriptions here should not be considered definitive. Health promotion programs in Blue Cross and Blue Shield Plans, as in industry, educational institutions, and communities generally, are being changed and expanded continuously in response to growing public demands for these services. So today's listing or inventory may be out of date tomorrow. Moreover, not all the new programs are reported or publicized in Plan communities, and some Plans have services or programs for their own employees that might readily be adapted for use by other employers if there should be evidence of interest among companies in the Plan community. So the only way an employer can be sure of having an accurate and current assessment of the preventive, fitness, education, and life-style behavior services that might be available through the Blue Cross and Blue Shield Plans in his area is by calling the Plans and asking. In most cases, any such inquiry should be directed to the vice president in charge of public relations. The officer with this responsibility would also know about any physical fitness or preventive activity, such as screening, that might be carried on by a Plan's health promotion, professional relations, or medical department.

Some clues about what these activities might be are listed here:[2]

- Blue Cross and Blue Shield of Connecticut sends trained teams to work sites to conduct onsite blood pressure screening. Plan personnel then train the company nurse to follow up treatment of hypertensive patients, check and encourage compliance with medication and diet regimens. The nurse is also trained to evaluate the effectiveness of the hypertension program for the group as a whole, including development of strategies for reaching more employees. For companies not having an occupational nurse on the staff, the Plan will find and train a suitable candidate. In any case, Plan personnel check periodically with company nurses and are available for consultation when problems occur.
- Blue Cross and Blue Shield of Delaware has a health education program designed for the work site, called "Safety, Health and Physical Education" (SHAPE). The program includes a health risk index assessment and offers courses in weight reduction, physical fitness, smoking cessation, stress management, and "defensive driving."
- Hospital Care Service Corporation (Chicago-based Blue Cross and Blue Shield Plan) conducts screenings and fitness awareness programs for employed groups. The Plan maintains screening task forces throughout the state.
- Blue Cross and Blue Shield of Greater New York is working with the 25,000 members and family members of the International Brotherhood of Teamsters Joint Council No. 16 in a series of Plan-sponsored seminars and workshops on improved nutrition, smoking cessation, weight loss, retirement planning, and other programs designed specifically to meet the needs of the Teamsters group. "This is the first time we've designed courses to meet the specific requirements of an organization," said New York Plan President Edwin R. Werner. "If the project proves successful we hope to expand our schedule of workshops so that all

Teamster locals can participate." The Teamster seminars and workshops are conducted at the health education center on the ground floor of the Plan's headquarters building in Manhattan, an unusual facility featuring health education exhibits and materials for the public.

- In order for groups to be eligible for participation in a new benefits reimbursement program developed by Blue Cross of Oregon, they must have fitness programs for employees in place. The Plan has a consultant who works with the groups to help them design and initiate programs matched to the particular needs of the groups. Lower absentee rates for participating groups are anticipated, if not guaranteed, by the "reimbursement benefit," which returns "unused benefit" checks at the end of the year to employees who have been absent less than the average rate for the group.
- Blue Shield of California's Stay Well Program pays subscribers to keep healthy. It was originally developed for employees of the Mendocino County Office of Education, where the program each year sets aside $500 for each insured subscriber; at the end of the year, any part of the $500 not used to pay for the employee's health benefits is put in a fund to be returned to the employee when he or she leaves or retires. The program also furnishes information that helps employees judge when medical care is needed. "We're not trying to discourage the use of needed medical care," Larry Parcell, Blue Shield vice president, explained. "We're only trying to discourage unnecessary care." At Mendocino, he said, when Stay Well was instituted, employees began to exercise, watch their diets, and "make positive efforts to get back the money."
- Blue Cross of Indiana has an extensive fitness program for Plan employees that is also available to other employed groups. Indiana is also participating with Blue Cross of Greater Philadelphia, Blue Cross of Virginia, and Blue Cross of Maryland, Inc., in a hypertension screening program that will include 15,000 members in the Federal Employees Health Benefits Program.

- Blue Cross and Blue Shield of Virginia have a health coordinator who visits employed groups to conduct seminars on exercise, nutrition, weight loss, stress management, and other subjects of interest to the groups. The consultant also assists groups in organizing and initiating their own employee fitness programs.
- Blue Cross and Blue Shield of Michigan has completed the second year of a three-year pilot project, "Go to Health," that involves 3,000 of the Plan's 5,000 employees. The program includes a computerized health risk assessment and appropriate education and counseling sessions for employees at high risk of cardiovascular disease. The Michigan program will be offered to employed groups following evaluation of experience with the pilot program. The Plan also has an extensive exhibit, "You and Your Health," providing information about cancer, heart disease, stress, smoking, and health care costs. The exhibit is available to employed groups and has been used extensively in education programs.
- Blue Cross and Blue Shield of Atlanta, GA, offers a health hazard appraisal program and computerized analysis of group risks to guide development of education, fitness, and prevention programs. When employees of Emory University were found to have high rates of use of medical services, Plan analysis of employee claims determined the reasons for physician visits and hospital admissions. Plan officials then worked with university personnel to create a health promotion program tailored to reduce the utilization rate for the group.
- Blue Cross of Western Iowa and South Dakota offers blood pressure testing services to employed groups.
- Blue Cross and Blue Shield of Maine has what it calls a Health Action Program for groups. The program features work site education in stress management, weight control, hypertension, smoking cessation, and fitness.
- Groups can borrow automated blood pressure testing machines from Blue Cross and Blue Shield of Massachusetts

for work site self-screening. The Plan also offers instruction in application of the Heimlich maneuver for choking victims, and the cardiopulmonary resuscitation "Heart Saver" instruction designed by the American Heart Association, teaching emergency procedures for saving victims of heart attack.

- Blue Cross of Northwest Ohio (Toledo) offers the Health Choice program in cooperation with other Ohio Plans and also provides cardiopulmonary resuscitation work site training and educational materials. The Plan has a work site cardiopulmonary screening procedure developed jointly with an area hospital.
- Blue Cross of Greater Philadelphia recently provided automated blood pressure machines at work sites as part of a one-month blood pressure public education program during which 27,000 blood pressure determinations were made. The Plan also offers instruction in the Heimlich maneuver for restaurant employees.

Nearly all Blue Cross and Blue Shield Plans, including those not mentioned here, carry on continuing, extensive public health education programs in their communities, with meetings, seminars, slide-films and movies, media campaigns, and materials for use by school, church, club, business, and other community groups. Many of the Plans also have traveling health exhibits and health fairs that can be presented at shopping malls and other population centers, including screening for hypertension, obesity, diabetes, and other health risks for as many people as may be interested. Educational materials and exhibits and screening services are also commonly offered to community groups, and while these activities may not be planned specifically for work site applications, for companies that know or can find out what their Plans are doing, it may be possible to arrange for use or adaptation of the educational services and materials for company employees.

The Health Engineer

A new approach to development of a company program for employee health and fitness that is being tried out by Blue Shield of California is the health engineer concept. The health engineer is a qualified consultant whose services enable a company to develop its own health promotion program for its own employees, using its own resources. The concept is new, but the health engineer is obviously similar to the safety engineer, now accepted as an effective resource in the design and management of accident prevention strategies. The health engineer would be a man or woman with a background in one of the health professions, though not necessarily in medicine, and in health promotion—ideally in public health, rehabilitation, exercise physiology, nutrition or physical therapy—and with some experience working in industry health programs. As in the case of Blue Shield of California, the health engineer would work with the health insurer's medical director and marketing people in planning positive health strategies for insured groups and persuading employers of the long-term wisdom of an investment in employee health and fitness.

Once the idea is accepted, the health engineer's function is to tailor the health promotion program to the company's interests and capabilities. The process begins with a comprehensive survey of the company that would include examination of the size, composition, and characteristics of the work force; the location, surroundings, and facilities afforded by the principal plants and offices; medical department, health care, and health insurance experience; absenteeism, turnover, and recruitment experience and costs; food and food service facilities and use; availability and accessibility of community resources such as hospitals, YMCAs, schools, parks, and others that might be used in health promotion activities. With all this information in hand, the health engineer can proceed with recommendations for indicated health education and fitness activities, or, if an employer's commitment permits, he or she might follow up first with an employee attitude study or perhaps actual screening for more detailed, precise health risk assessment. In any case, the health engineer's recommendations can be presented in the form of a proposal for company action, including an explanation in full of the costs as well as the potential benefits to the firm. "Once the firm has decided to initiate a positive health program for its employees, either company-wide or segmented, the health engineer can serve either as consultant or implementor," said Larry Parcell, vice president of Blue Shield of California, who—working with the University of California, San Diego—developed the health engineer concept.[3] The health engineer could serve as the onsite supervisor of the program, Parcell suggested, or he or she might furnish supervisory staff as needed, or train company personnel for these functions. As proposed by Parcell, the health engineer would be employed or retained by the health insurer, at least up to the point of implementation of the program. "By providing these services the carrier would demonstrate innovativeness in its approach to health and health financing, as well as performing a useful service to the firms in the area," he said. "It also would be providing something many major U.S. companies have declared that they want—one-stop shopping for as many as possible of their health needs."

The first health engineer program—a pilot project—was started in September 1981 utilizing the services of a pioneering health promotion organization, Life Management Group of La Jolla, CA. In addition to helping implement the program, Management Group also served as consultant in development of the concept of the health engineer.

Put Some Pep in Your Step

Among the other health promotion activities Blue Cross and Blue Shield Plans have initiated to increase interest in health and fitness in their communities, many of the Plans have joined in an effort to encourage a form of exercise that physical fitness authorities consider to be a superb conditioner—one that requires no resources of money, equipment, personnel, or even very much time, and is as readily available to employees everywhere as the nearest door or flight of stairs: walking. "Not everyone can be a jogger, but just about everyone can walk regularly," says a Blue Cross and Blue Shield Associations report, "and the benefits are enormous: a healthier heart, a healthier body, and a healthier frame of mind." The associations and the national YMCA organization are cooperating in a program aimed at publicizing the health benefits of walking. Called "Put Some Pep in Your Step," the program is designed to show community organizations how they can raise public consciousness of walking as an easy, enjoyable, and healthful exercise. Blue Cross and Blue Shield Plans and YMCAs taking part in the program distribute a booklet on walking that was prepared for the President's Council on Physical Fitness and Sports. "When done briskly on a regular schedule," the booklet says, "walking can improve the body's ability to consume oxygen during exertion, lower the resting heart rate, reduce blood pressure, and increase the efficiency of the heart and lungs. It also helps burn excess calories. Since obesity and high blood pressure are among the leading risk factors for heart attack and stroke, walking offers protection against two of our major killers."[4]

The thing that makes walking a healthful, beneficial exercise instead of simply a pastime or a chore, the President's Council emphasizes, is largely a matter of pace and distance. "When you're walking for exercise, you don't saunter, stroll, or shuffle," the booklet explains. "You move out at a steady clip that is brisk enough to make your heart beat faster and cause you to breathe more deeply." In some weight loss and conditioning studies, the Council reported, this kind of walking actually proved to be more effective than running or other touted fitness exercises, "because it's virtually injury-free and has the lowest dropout rate of any form of exercise." And "walking appears to have a substantial psychological payoff. Beginning walkers almost invariably report that they feel better and sleep better, and that their mental outlook improves. Walking can also exert a favorable influence on personal habits. For example, smokers who begin walking often cut down or quit. There are two reasons for this. One, it is difficult to exercise vigorously if you smoke, and two, better physical condition encourages a desire to improve other aspects of one's life."

Fitness experts recommend that beginners walk for 20 minutes four or five times a week at a pace that is brisk but not tiring, and they warn that persons who have been inactive for years should guard against overdoing, adjusting pace and distance to avoid strain and produce a feeling of well-being. "You should be able to carry on a conversation while walking," the Council advises. "If you're too breathless to talk, you're going too fast. If you develop dizziness, pain, nausea or any other unusual symptom, slow down or stop. If the problem persists, see your physician before walking again. Your objective should be to improve your own performance, not to walk farther or faster than someone else. The most important thing is simply to set aside part of each day and walk. No matter what your age or condition, it's a practice that can make you healthier and happier."

If direction and assistance from a hospital, Blue Cross Plan, YMCA, or other community source isn't available, any employer can start his own fitness program by encouraging employees to walk, using whatever means of communication are at hand within the company to stress the benefits and make suggestions. Employees can be urged to walk part of the way to and from work, for example, to walk in groups at lunch time—if the work site neighborhood is safe and convenient—and to use stairs instead of elevators whenever possible. And if the fitness advisers to the President's Council are right, walkers can be expected to become interested in other means of taking better care of themselves. Smoking on the job will become less noticeable; fruits and vegetables will appear in lunchrooms and snack bars. Managers who are alert about what's happening will see the signs of growing interest and look around for opportunities; if they haven't been interested in investigating community health promotion resources before, they'll find out now what local hospitals and Blue Cross and Blue Shield Plans have to offer. Or there may be a nearby YMCA, or school, or college with facilities that interested employees could use for more organized exercise groups—perhaps with staff members who might be glad to take on a class in nutrition, or weight control, or smoking cessation. It's happened just that way in some companies, in some communities. And once an activity has taken hold, and results in improved employee performance and morale are observable among the participants, it's not hard to take the next step and move on to an organized health risk assessment and education program for everybody. When you start with what you know you can afford, you find that you can afford what you know you ought to have.

Another way of testing the wellness water before taking the plunge is offered by the Wilson Learning Corporation of Minneapolis, a company that got into the wellness business as a means of supporting its original functions dealing with employee attitude studies and initiating all kinds of corporate training programs.

"How can we help performers become more versatile, more effective at learning, more able to translate learning into productivity?" the company asked itself. "How can we help them drop their self-limiting beliefs about what they can do, and expand their notions of what is possible? As we started addressing the problem in this way, we began to see the importance of stress management, of fitness, of nutrition, of self-esteem, of the human spirit—in short, of wellness."[5]

Wilson Wellness

The Wilson concept is that the mind and spirit, as well as the body, are essential components of wellness. "Unconscious beliefs play a major role in limiting our ability to respond flexibly and effectively to stressful events," says a brochure describing "Wilson Wellness," as the programs are called. "Becoming aware of our automatic reactions and unconscious beliefs and their limiting influence is a big step toward quality of performance, health, and satisfaction. Recognition of personal responsibility and choice; the capacity for relaxed concentration, for spontaneity and creativity; acceptance and appreciation of self and others—all are key skills of the mind that each of us can learn and make work for us in every area of life." And, completing what has come to be called a wholistic view of health, the brochure adds, "We are also discovering that many 'skills' traditionally identified with religion are as good for the psyche as for the soul: meditation and contemplation, the appreciation of beauty, the experience of joy and wonder, receptivity, acceptance, service, and love."

The Wilson goals of joy, love, and creativity, as well as knowledge and fitness, are all sought in Wilson Wellness seminars, which may take from one to three days and include either executives or supervisory personnel or rank and file employees, or all of these, depending on the range and depth of the client company's interest. The seminar is a learning program designed to be a beginning, not an end. But the lectures, movies, demonstrations, and other learning materials pretty much cover the wellness universe—nutrition, fitness, exercise, lifestyle, and, as participants are told, "a personal action plan and a support system to help you achieve your goals for physical, emotional, and spiritual well being."

Trial Run

For companies that are convinced, as many are, that a one-shot program, however well-loaded the shot, isn't likely to change many lives for very long, Wilson Learning offers continuing wellness programs aimed at securing a firmer grip on the seminar goals. Meanwhile, Wilson's wellness staff, headed by John N. Williamson, Ed.D., is putting the finishing touches on a broader and deeper trial run at wellness for companies that want something more than a seminar but are still not ready for a two- or three-year commitment—something in between a toe in the water and a swan dive. The new program is called "How to Get Wellness Started," and is described by Dr. Williamson as "both an essential, state-of-the-art overview of corporate wellness issues and a practical guide to educate selected employee groups in the skills of engaging large numbers of employees in activities assuring a high degree of worker participation."[6] How to Get Wellness Started includes the Wilson Wellness seminar and adds "a structured, two-day classroom experience supplemented by a variety of preseminar materials, classroom video support, and post-seminar action steps and other support materials," according to Williamson.

The intention of How to Get Wellness Started is to guide the select employee group—and through them the work unit, division or company—in assessing needs, priorities, and resources, reviewing the range of available options, and initiating an action plan, Williamson explained. "How to Get Wellness Started provides a theory base and initial activities pointing toward a self-sustaining, evolving health promotion program that is an expression of the employees themselves and within the resources of the organization," he added. The idea is that the group will become a company wellness task force that can assemble and examine baseline health data, and then, if the management decision is to go forward with development of a companywide wellness program, task force members will be prepared to establish goals, work out priorities, and undertake the process of organizing employee participation, calling on Wilson Learning consulting services as needed. Williamson estimates that the cost of How to Get Wellness Started for a company, division, or department of about 300 or 400 employees would be approximately $5,000. The program thus offers a comprehensive view of wellness, he said, but "it does not require commitment to any comprehensive solutions. The format is flexible and points to a variety of options and detours that can be taken from the essentially simple track suggested."

Chapter Nine
KEEPING SCORE: THE OUTLOOK FOR WELLNESS

"Physical fitness is maintained only through a normal regimen of healthful living. There is ample evidence of the effects of living on health." So wrote C. E. Turner, Sc.D., professor of public health at the Massachusetts Institute of Technology, in a textbook on personal and community health published in 1925. The only evidence Professor Turner mentioned was the effect of athletic and military training and dietary treatment of malnourished children, and he didn't feel called on to specify what this effect was, presumably considering that everybody would accept the truth of the proposition on the face of it, much as we do today. But the normal regimen of healthful living hasn't changed over the 50 years from Turner to Breslow. Turner was concerned about a few things Breslow didn't include but would probably agree with: four glasses of water as well as three meals a day, for example, a daily bath, and dental hygiene. Breslow was explicit about alcohol and smoking, which weren't included in Turner's normal regimen, but not because he wasn't looking. "The use of alcohol has a marked effect on longevity," he said. "Statistics from life insurance companies show that on an average total abstainers live longer than non-abstainers: there is evidence that alcohol even when used in apparent moderation is a factor in shortening the average length of life."[1] And back in the days when pipes and cigars were more visible than cigarettes, and women who smoked stayed out of sight, Turner was stern about tobacco. "The excessive use of tobacco tends to increase the heart rate," he declared. "Motor disturbances which are evident through tremor and loss of control of the finer muscles result in reduced skill in manipulation. Growth is definitely retarded under the influence of tobacco. Until growth is completed, tobacco should not be indulged in."

If we knew so much 50 years ago, why has it taken so long for good health habits to become the generally accepted mode of living, instead of a seemingly new discovery? And why don't we have more conclusive evidence now of the results of healthful living?

These are good questions, and there aren't any easy answers. Long before our time, Turner may have had a sense of the influences and distractions that lead us into bad habits. "The struggle for social and material progress crowds out consideration for our physical well-being," he wrote. That was years before Hans Selye invented stress and mapped its physiologic booby traps. But Turner, with an instinct for what was needed, urged his readers to seek a "general attitude of cheerfulness and confidence in relation to life" and "freedom from persistent worry and anxiety." Life is a process of continually adapting, fitting, and adjusting oneself to one's surroundings, he advised, and his rules included one day a week for resting and two afternoons or evenings a week for "nonprofessional activity"—meaning, presumably, anything unrelated to work.

These are not bad ideas today, and Selye would unquestionably applaud, but they are out of step with the modern world of workaholic executives. However, there are signs that the business world may be catching up with Dr. Turner's concept. Corporations report that many young executives are turning their backs on promotions that mean transfers to big cities and less time for families, and the growing number of companies that are investing in fitness and health promotion programs for employees may discover along the way that overnight and weekend assignments and pressure for productivity are part of the problem, not its solution.

The signs that good health pays more than it costs are multiplying, too, although the cause-and-effect relationships are still more assumed than explicit, and are not readily translated into dollars. For example, the incidence of coronary heart disease has been declining—22 percent in the past decade, according to a 1980 report from the Surgeon General of the Public Health Service. It is not possible to isolate and quantify the causes, which are obviously complex and interrelated, but it is not sensible to ignore them, either. The National High Blood Pressure Education Campaign that began in 1973 and continued throughout the decade is certainly one cause. The decline in smoking, especially among adult males, is another. A shift in the nation's eating habits toward fewer animal fats and more fresh fruits and vegetables in the national diet is still another. And in the context of this report it would be witless not to cite what has been called the "fitness mania" that has uncounted millions out there every day jogging or running, riding bicycles to work or in their basements, doing pushups—some impelled by social pressure, to be sure, but for the most part convinced that they are improving cardiopulmonary performance and, as they devoutly testify, feeling better.

The Stanford Heart Program

Fortunately for industry, there are more explicit results than feeling better. For the past several years, John W. Farquhar, M.D., professor of medicine and director of the Stanford Heart Disease Program at Stanford University Medical Center in Palo Alto, CA, and Nathan Maccoby, Ph.D., director of Stanford's Institute for Communication Research, have been conducting a continuing evaluation of the effectiveness of health education and instruction in three comparable California communities within a few miles of the Medical Center. In sample screenings of the community populations, baselines of health attitudes and behavior were established and segments of the populations at risk for cardiovascular disease were identified. In one of the communities a mass media heart disease education campaign was introduced including radio and television advertising, doctors' and dietitians' newspaper columns, cookbooks, booklets, and other special educational materials and programs. Also in this community, the cardiovascular risk population received face-to-face instruction consisting of 10 intensive one- to three-hour sessions in group meetings and at home. A second community in the study received the mass media education but no face-to-face instruction. No education or counseling of any kind was offered in the third community.

At the end of each of the first three years of the study, all three communities were resurveyed and retested. In the first community, the education and instruction programs had produced significant declines in cardiovascular risk for the population as a whole, and for the high-risk segment, as revealed in blood pressure determinations, incidence of heart disease, and in health attitudes and behavior. The overall population of the education-only community showed a similar decline in cardiovascular risk at the end of the first year, but by the end of the second year the risks were about the same in both communities for the overall populations. The first community, however, had retained its risk advantage in the segment of the population that had received the special instruction. No change at all in risk or behavior was noted in the third community. "The net effect of the changes in smoking, plasma cholesterol, and blood pressure was to create a rather surprisingly large degree of reduction in the multiple logistic of risk," Dr. Farquhar said, commenting on the three-year result. "This change as compared to control (the third community) was approximately 30 percent for the intensively instructed group and 24 percent for the media-only participants. The examined cohorts in the control community showed modest increases in blood pressure and cholesterol. This finding was not unexpected and was quite analogous to the experience of many prospective epidemiologic studies which have regularly shown an age-related increase in risk factors. . . . From this study we conclude that a carefully planned treatment program that relies almost entirely on the mediated instruction of mail and telephone follow-up can result in significant changes in risk factors that are maintained over long periods of time. These results are gratifying, and they offer further encouragement that mediated methods of health education can work in different settings," Dr. Farquhar observed.[2]

To further validate the usefulness of the mass media approach, the investigators are now engaged in a five-community study that is planned to extend over seven years, Dr. Maccoby reported. "The media components of this new study will include the electronic mass media, the printed mass media, school programs for age range 8 to 18 years, employee programs, adult evening classes, etc." he told a conference on health education sponsored by the Health Insurance Association of America (HIAA) and the American Council of Life Insurance. "Community organizations are being helped to carry out programs of face-to-face instruction so that health education can be self-maintaining. This time, we will be studying not only change in risk but also change in morbidity and mortality. In this way we hope to demonstrate definitively that communitywide educational methods can be an effective tool in changing behavior so as to help maintain health, and that these methods can be generally adapted to communities everywhere."[3]

Corporations that have made major investments in employee health promotion programs are naturally concerned about results, and most of them, like Control Data Corporation, have set up computerized data systems designed to produce outcome measurements. As in the case of the Stanford Heart Disease Program, however, early results can show only behavior changes and statistical risk reduction. These are important indicators, but actual savings depend for the most part on decreased morbidity and mortality that are determinable only over longer periods.

Dr. Ernst Wynder, whose American Health Foundation in New York offers comprehensive screening and health education services for industry groups, likes to tell business executive audiences that the entire cost of such a program for a group of 1,500 employees for a whole year would be returned to the employer if a single coronary bypass operation were avoided as a result—a statistic that always impresses members of the audience who don't stop to consider that they would never know whether a coronary bypass had been avoided or not. The only valid outcome measurements in health education are those that trace the incidence of illness and injury in groups that are large enough over periods that are long enough to translate changes in behavior into changes in the quantity and cost of medical services consumed.

Results at Campbell Soup

The Campbell Soup Company has had an employee health promotion program since 1968, Dr. Roland F. Wear, Jr., corporate medical director, reported at the HIAA conference. The nature of the screening and content of the education program were shaped by analysis of employee health records and statistics for working populations, he said, which "helped us to focus on the most common causes of death, disability, and hospitalization of working people and to consider those with cost-effective methods of screening, intervention, and prevention." For example, "in February 1968, our atherosclerosis risk factors were looked at and interventions were tried. After several years of experience, it was obvious to us that we obtained our best sustained results in hypertension control, rather than in correction of obesity, elevated serum lipids, or smoking. We decided to continue all our efforts, but to intensify those in hypertension control. We attempted to screen our entire population and to follow all hypertensives and borderline hypertensives. The persuasion powers of an interested nurse appeared to be our most potent weapon. Information handouts and referrals to physicians skilled in hypertension management were also important elements in our program."[4]

A recent survey of all plant nurses indicated that employee awareness of the presence of hypertension had increased from 50 to 99 percent, Dr. Wear reported, and from 60 to as many as 90 percent of the hypertensives were considered to be on adequate treatment. The annual costs of inhouse hypertension treatment were estimated to be $100 per patient for those treated by company doctors and nurses, compared with private care at an estimated $300. The program was probably not cost-effective in the first years, Dr. Wear acknowledged, "but the more severe the hypertension or the older the population, the sooner and greater the benefits. In the age group 55 to 65 years, 100 employees will include about 25 hypertensives with annual treatment costs of $7,500 for private care. Ten-year costs would be $75,000 for private care. In a population of 100 persons aged 55, we would expect up to four strokes with one death prior to age 65. Almost three of four strokes in this age group could possibly be prevented by hypertension treatment; thus $75,000 would be spent to save up to $130,000, which is the approximate cost of three strokes in terms of hospitalization, rehabilitation, long-term disability, lost wages, etc. The incidence of hypertensive heart failure and other complications could also be reduced by treatment. We have attempted to develop some programs that cost no more than they eventually save."

The desirability of the work place as a site for treatment as well as detection of hypertension was established a few years ago in studies of members of the United Storeworkers Union employed at Gimbel's and Bloomingdale's department stores in New York City. Eighty to 90 percent of all employees volunteered to have their blood pressures taken, and the 10 to 15 percent of members identified as hypertensives were offered the opportunity to get all subsequent care at the stores, Michael H. Alderman, M.D., professor of medicine at Cornell University Medical College, who directed the study, reported. Seventy-five percent of them chose work place treatment initially, Dr. Alderman said, and more joined later at the urging of satisfied participating colleagues.

United Storeworkers Union

Four years later, Dr. Alderman reported that about 80 percent of all those who started on the program had achieved and maintained satisfactory control of blood pressure. Significantly, about the same result had been obtained for a group of 300 patients who joined the work place treatment group "after failing to achieve blood pressure control in conventional treatment settings," he added. Since the content of treatment is the same for both groups, the advantage of the work place, obviously, is in the supervision and follow-up afforded at work and the convenience of work place treatment—features that are missing when the patient must motivate himself to make and keep doctors' appointments and comply with medication and dietary treatment regimens. Dr. Alderman is convinced that compliance, not detection, is the answer. "Probably as a result of the attention that has been lavished on the problem of hypertension, it can no longer be described as an unknown disease," he said. "The crux of the problem has dramatically shifted in recent years. Universal detection has either been or is about to be achieved. But this very success has highlighted the fundamental problem of the continuing weakness of conventional approaches to the delivery of antihypertensive care."[5]

Another hypertensive specialist, Frank A. Finnerty, Jr.,

M.D., of Washington DC, explained the phenomenon. "It has been my experience, both in private practice and in the clinic, that patients will not remain under medical care and on medication unless they are properly motivated," he said. "Such motivation can result only from a good doctor-patient relationship. Education in the absence of such a relationship is worthless. The successful long-term treatment of hypertension is dependent on and is the result of a good doctor- or nurse-patient relationship. Once the hypertensive patient has been initially evaluated and placed on a therapeutic regimen by a physician and has reached a status quo situation, he can ideally be followed by a nurse or health assistant working under a nurse."[6]

Dr. Alderman agreed. In the storeworkers program, "treatment was initiated by the supervising physician after thorough physical and appropriate laboratory examination had established that the employee met the criteria for entry. Nurses were primarily responsible for follow-up visits. A careful record system ensured close physician supervision; the physician also attended each work site regularly for consultation, review of problems, and patient examination when necessary, and re-evaluated each patient in an annual examination. . . . This experience sustains the belief that detection and treatment of high blood pressure at the work site offers a safe, highly acceptable, and very effective method of blood pressure control for very large numbers of persons."

Lawrence W. Green, Dr. P.H., who headed the division of health education at the Johns Hopkins School of Hygiene and Public Health and was later director of the Office of Health Information, Health Promotion, and Physical Fitness in the U.S. Department of Health and Human Services, has explained some of the difficulties of making precise measurements of the results of these programs. Scientific rigor requires strict adherence to a protocol specifying experimental conditions and procedural detail, he said, "but the attempt to maintain such rigorously defined procedures often results in sterile, perfunctory, or routinized performance which is not sufficiently adapted to emerging circumstances to be significant in its impact. Thus we end up sometimes with rigorously defined but trivial interventions and at other times with significant interventions that are too vaguely defined to be replicated."[7] The purpose of the research protocols Dr. Green referred to is to exclude other variables that might influence behavior so the impact of the health education activity, or intervention, as he described it in research terms, can be isolated for precise measurement. If people change their eating and drinking habits during or following a nutrition education sequence, for example, in order to credit the change wholly to the education sequence we should have to exclude the possibility that other influences, such as family pressure, or perhaps an educational television series, might have been at work at the same time. Since research conditions can rarely be duplicated outside the laboratory, it's a good thing that companies are more interested in the practical outcomes than they are in the purity of the procedural details.

This is not to say that research in the effectiveness of health education is not needed and useful. The more that is known about the impact of such programs, the more likely their content and presentation are to have desirable and lasting results. For example, it is because of research studies by Dr. Green, Dr. Maccoby, and others that many of the corporate health education programs call for periodic repetition, or, as Control Data does, an initial commitment for three years. "We know a great deal about learning curves, and memory curves, and the processes of forgetting and back-sliding," Dr. Green said. "We know that reinforcement is as important to education as booster shots are to sustained immunization," he emphasized.

Kimberly-Clark

In addition to providing the needed booster effect, annual rescreening and repetition increases participation. One of the most comprehensive of the corporate projects, Kimberly-Clark Corporation's Health Management Program, was offered for the first time in 1977 and has had steady growth since then in use of the company's extensive exercise facility, in the numbers of employees screened, and in attendance at classes in everything from "Saving Your Back" to "How to Talk With Your Doctor"—a useful innovation in the Kimberly-Clark education sequence. The company also has an Employee Assistance Program offering counseling for employees with alcohol or drug or other special problems.

"Measurement of outcomes such as cost-benefit results, risk factor and absenteeism reduction, and program adherence are all important for the future," Robert E. Dedmon, M.D., vice president, commented. "Final success will probably be achieved only if these programs evolve into what is truly a cultural change in the eating, exercise, and other habits of the population." In a 1980 report, Dedmon said screening had identified hypertension, obesity, hyperlipidemia, and other health problems in many employees. Significant heart disease requiring surgery was revealed in several employees, and eight cases of undetected cancer were discovered. "The rehabilitation success rate for employees with chemical dependency problems with respect to job performance is 65 percent," he added. "Absenteeism and accident data for a group of employees participating in the employee assistance program showed a 43 percent reduction in absenteeism and a 70 percent reduction in accidents for a one-year period following treatment, when compared with a one-year period prior to treatment for chemical dependency."[8]

Attempts to measure the results of the fitness components of health promotion programs are even more difficult than those aimed at determining the effectiveness of screening and education. Since there are fewer objective measurements that can be made, evaluation of fitness programs depends heavily on the participants' own assessments of what has happened, as in the study conducted at the Washington, DC headquarters of the National Aeronautics and Space Administration, reported in Chapter 2. In that case, the perceptions of the NASA subjects about their own health were verified by objective measurements of blood pressure, pulmonary function, heart rate, body fat, and other indicators.

Results in Canada

One of the most extensive studies of the effects of fitness programs was conducted in a group of 1,200 insurance company employees in Toronto, Canada, in 1977 and 1978. In the Employee Fitness and Lifestyle Project, as it was called, employees of the Canada Life Assurance Company and the North American Life Assurance Company were tested and studied over an eight-month period during which the Canada Life employees participated in a fitness program that included regular exercises and a variety of health education and lifestyle classes and activities. These employees were divided and tested in separate groups of "high adherents" to the exercise and instruction regimens, "low adherents," "drop-outs," and "nonparticipants." The North American Life employees were not exposed to any kind of fitness activity or instruction and served as the control group.

Over the period of the experiment, participants showed moderate improvements in such measurements as body weight, body fat, muscular flexibility, and cardiorespiratory fitness, it was reported. However, "the overall fitness performance of the program participants did not improve as dramatically as may have been expected," the report said. "The short class periods of 30 minutes did not allow an extended aerobic workout during the class. In order to include a variety of exercises aimed at improving muscle tone and flexibility as well as cardiorespiratory fitness, the aerobic portion of each class built up to a maximum of only 15 to 17 minutes. Most of the exercise classes were very low key initially, due to the low fitness levels of participants. Also, several classes included older persons, overweight individuals, and persons who were in an organized program for the first time. This forced the instructors to build up the intensity of the workout very gradually."[9]

Improvements in absentee rates were something else. The overall absentee rate for both the experimental and control companies declined over the period of the study because of an epidemic of influenza that occurred in Toronto in December 1977, the "going in" period of the experiment. When the subgroups at the experimental company were examined separately, however, the high-adherence group of regular exercisers at Canada Life showed a 42 percent decline in absenteeism over the study period—22 percent greater than the decline for employees of both companies had been. The decline in absenteeism was estimated to have been 2.5 days per year per employee, or $175,000 a year in salary savings at average salary rates for the 1,400 employees of Canada Life.

Productivity assessments were made by objective output measurements in some departments but had to rely on supervisors' subjective judgments in many cases. Productivity increases over the study period varied wildly among the departments, as it turned out; interestingly, the moderate overall increase in productivity in Canada Life was duplicated at the control company, where no fitness activity had been conducted. Commenting on this result, the report said that "the productivity levels at the experimental company were initially very high, not leaving much room for improvement regardless of the type of intervention used. The effects of onsite fitness programs on productivity need to be explored more thoroughly. Ideally, a similar experiment should be conducted using an industrial setting with highly accurate methods of assessing productivity."

Employees in all the research subgroups were asked to record their visits to the occupational health units of their companies. Visits to the company nurse and doctor were tabulated separately in an attempt to determine if participation had an effect on the frequency of minor medical complaints. The tabulations indicated that both male and female high adherents tended to have fewer visits to the nurse than low adherents, nonparticipants, and controls. Similarly, visits to the company doctor were less frequent for the high adherent program participants, it was reported. "The most dramatic difference was between female high adherents and the female control group," the report said, "with the latter group reporting an average of .4 visits per person compared to .1 for the high adherents. If this can be considered indicative of the typical working population, these results seem to suggest fewer health problems amongst regular participants in fitness programs."

On the purely subjective questionnaire measures of such things as job satisfaction, morale, and attitudes toward activity and health, 47 percent of the employees who participated in the fitness program reported that they were "more alert, had better rapport with co-workers and supervisors, and generally enjoyed their work more since starting the program. On the personal part of the questionnaire," the report continued, "63 percent of the program members indicated that they felt more relaxed, patient, and less tired during the day since their involvement in the fitness classes. Thirty-two percent expressed no change in personal feelings, while 5 percent indicated other changes but did not specify them. Although these questionnaires were subjective in nature, there is no doubt that the fitness and lifestyle program at the experimental company had positive effects on the program participants, other than simply improvement in fitness," the report concluded. "The most positive results were in improved general fitness amongst program participants, substantial decrease in absenteeism and employee turnover, and improvement in the morale and attitudes of participants with respect to both their work environment and personal health. In conclusion, it seems fair to say that certain benefits will accrue to the company and the individual employee if a fitness and lifestyle program is implemented."

The Relaxation Response

In another work place assessment that involved both perceived and measured outcomes, an experimental group of employees at the Converse Rubber Company tested the effectiveness of the "relaxation response" technique developed by Herbert Benson, M.D., professor of medicine at Harvard Medical School. In the experiment, one of the randomly selected employee groups received instruction and practiced Dr. Benson's technique, which includes: 1) a comfortable position in a quiet environment; 2) muscular relaxation; 3) conscious deep breathing; 4) a passive attitude supported by a mental device such as repetition of a single word or thought. Another group was instructed simply to sit quietly for comparable periods—10 to 20 minutes twice a day. Members of both groups and control groups of selected and volunteer participants all followed prescribed methods of assessing such outcomes as symptoms, illnesses, performance, sociability, and happiness.

Analyzing the reports, Dr. Benson and his associates found significant differences among the groups in blood pressure measurements, behaviors, and perceptions and concluded that the relaxation instruction resulted in appreciable benefits for the instructed group in comparison with the others. "The relaxation response training involved a minute investment of time and personnel to achieve the desired results in comparison with other programs of stress control which some employers have purchased," they said.[10] Commenting on the result of this and other experiments, Dr. Benson remarked that if high blood pressure can be alleviated by behavioral means alone, "its cause may also lie in a behavioral mechanism. By far the most appealing use of the relaxation response in relation to hypertension lies in its preventive aspects. To establish the preventive role of the relaxation response in hypertension, it is necessary to conduct large, expensive, and difficult investigations that often take many years to complete."[11]

Dr. Benson's investigations are continuing, and his emphasis on long-term evaluation of health promotion outcomes echoes the emphasis on long-term repetition and support of lifestyle education and behavior change that has been apparent in all the health promotion programs reported here.

Dr. Robert Allen of the Human Resources Institute explained why sustained effort is an essential element of successful fitness and health education programs. "People are living in a sea of negative norms," he said. "Their smallest cultural units, such as their families or work groups, can be seen as illness cultures full of negative norms that work against their best intentions. And these smaller units are surrounded by larger ones, consisting of our organizations and communities, and the even larger ones of state, nation, and world—most of which are illness cultures, full of large numbers of negative health norms. Consider how much more difficult it is for a person to stop drinking, for example, when his family and friends continually offer him drinks, his company celebrates promotions with cocktail parties, his community has many liquor stores, national advertising relates alcohol consumption with popularity and virility, and the western world has norms about individual heroics that make him think he should fight his problems by himself!"[12]

Healthy Conventions

A common if not universal negative norm in modern life is the corporate or industry or professional convention, where participants sit still all day and stand up drinking all night, or so it seems, and then return to their usual lives and tasks—presumably renewed by what they have learned but more often than not paralyzed by what they have experienced. The corporation that initiates a fitness program for its personnel but continues its typical convention routine is, in effect, stepping on the gas and applying the brakes at the same time, a practice that is likely to kill the engine.

Convention fashions are changing, to be sure. There are smoking and nonsmoking sections in many meeting rooms today, for example, and the sternum-to-sternum shouting match, or cocktail party, is depleted or foreshortened on occasion by the absence or early departure of the dedicated early morning joggers. But conventions serve a useful purpose, and there is no inviolable law that says they must also be hotbeds of unhealthful behavior. Instead, they could afford the corporation with a real interest and investment in the health of its people an opportunity to innovate. "Why not structure an unconventional convention?" asks Martin L. Collis, Ph.D., professor of physical education at the University of Victoria, in Canada. Among his suggestions for true believers are: 1) no smoking in the main hall; 2) morning and afternoon miniexercise breaks; 3) nutrition breaks with juice or fresh fruits and bran muffins as well as coffee; 4) planned walks at lunch time; 5) special presentations on health-related topics; 6) carefully selected nutritious meals; 7) exercise opportunities before dinner; and 8) maybe a vigorous dance in the evening.

"There is nothing in Roberts' Rules or the 'How to Run a Conference Manual' that says the participants must be sedentary," Collis concluded. "Nowhere is it stated that people should not be allowed to have fun and derive some fitness benefits from their conference participation. Most hotels will respond favorably to requests for menu changes and other details. Remember, the conference is being run for the benefit of the conferees, not the convenience of the hotel staff."[13]

But the cultural norms that surround us are susceptible to change, Allen insists. "This fact is both hopeful and scary," he warned. "It is hopeful in that it points out that many of the tragic norms of the past, such as racism, poverty, and mistreatment of women, can be overcome much more rapidly than most of us had thought. It is scary in that, as we have seen in our own lifetime in Nazi Germany and the My Lais of Vietnam, it is possible for people to set aside positive cultural values in favor of murder, death, and destruction. Based on our experience in helping people themselves challenge such problems as migrancy, delinquency, drug abuse, racism, and corporate ineffectiveness, we believe that cultures can be changed from a threat to life to a protector of life.

We believe that people do not have to continue to live in a negative health culture. We also feel that it will be only through such cultural changes that the massive body of work in health hazard appraisal and prospective medicine can be converted into greater gains in the length and quality of individual lives."[14]

In Dr. Allen's Lifegain, Control Data's StayWell, Skokie Valley's Good Health, Kimberly-Clark's Health Management, Blue Cross and Blue Shield of North Carolina's EASE, and others that have been reported here and elsewhere, it has been demonstrated again and again that there are a few indispensable components of a successful company health promotion program. These may be described in different terms and given different weights, but it is clear that whether the company is large or small and whether the program is ambitious or modest, if any of the parts is missing, the effort is going to fall short of its goals, and the program probably isn't going to be around for very long. The hard bones and long muscles of company health promotion are:

1 **Commitment.** Top management must be in the act from the beginning, with the understanding that it is important, that it will require an investment, and that it will be months, if not years, before the results can be measured in dollars.

2 **Knowledge.** The program must be designed to fit these particular employees in this particular place or these places, and this requires an initial assessment of the health status, or culture, of the employee population.

3 **Leadership.** The person or persons in charge—the managers or consultants or whoever it is who sets goals and initiates programs—must have been around this track before. These are complex activities, and even the do-it-yourself change agent needs a blueprint.

But hard bones and long muscles don't make a whole body. It takes flesh and blood and vital organs to make the structure come alive, and these essential parts of the company health promotion program are supplied by:

4 **Involvement.** Anybody can start and manage a health program for the dedicated jocks and joggers, who are already fit and are going to stay that way no matter what. The trick is to

involve a strong cadre of rank-and-file employees initially and direct their enthusiasm to engage new recruits.

5 **Support.** Whether it's dieting or not smoking or pushups or stress management, the key ingredient of health promotion isn't the content or the instruction so much as it is the peer support, fueled by the fine art of identifying and encouraging group activity leaders.

6 **Renewal.** Changing lifestyles isn't accomplished in a few weeks or months, and changing the company's health culture from negative to positive takes even longer, so renewal—new participants and booster shots for old participants—is imperative, and the foundation for renewal is evaluation of what has been accomplished.

New Cultural Patterns

The trend of events in the national health care economy over the past three decades has made it clear that we can't go on eating, drinking, smoking, working, driving, and worrying ourselves into ill health and then paying the health care system to repair the damage. The brakes on system expense that are now being applied may provide temporary relief, but the long-term solution obviously lies in the new emphasis on prevention, with business and industry leading the way. As companies change their health behavior norms from negative to positive, the new cultural patterns will spread to families and communities and beyond. In time, the national health culture will begin to shift; bad health habits will give way to good health habits, and our children or grandchildren will live longer and feel better and be relieved of the burden and worry of uncontrollable health care expense.

If it doesn't happen that way, if for want of enough companies with the commitment and knowledge and leadership and involvement and support and renewal it takes to change the health behavior culture of business and industry—the driving force of the American economy—the emphasis on prevention that will come inevitably could come from another direction. What that direction might be was suggested to a reporter who visited hospitals and public health stations in the Soviet Union a few years ago with a group of American public health physicians. At the Institute for

Public Health in Moscow, the group heard a lecture by a professor of public health who described the USSR's industrial health system. At every place of employment, he said, a prescribed staff of physicians and nurses—so many of each per hundred or thousand employees—is on the premises at all times. At specified intervals, all employees are given complete medical examinations, the results of which are forwarded to the employees' neighborhood clinics, where they are then required to report for whatever treatment or regimen is indicated.

At the end of the lecture the professor wanted to know if there were any questions, and one of the American physicians asked: "What do you do if they don't all do what they're supposed to do?" The professor seemed puzzled and asked to have the question repeated, and the doctor asked again how noncompliance was handled. The professor thought for a moment, then he smiled and answered the question.

"It never happens," he said.

Bankable Assets

More suited to the American style is the example provided by Jess Bell, head of Bonnie Bell Cosmetics, whose personal experience demonstrates how commitment, leadership, and involvement can change the lives of an entire group. Bell likes to tell the story of how it came about, and often does. "I had reached bottom," he said of his own life, looking back to 1972. "My marriage was on the rocks, I was 30 or 40 pounds overweight, and I was looking at the world through the bottom of a martini glass."[15] The business wasn't in such great shape, either, Bell said, and he attributed his gallimaufry of problems to a combination of executive stress and "the middle forties syndrome."

Then he stopped drinking and started running, Bell related. Before long he was reunited with his wife, and within two years he began sharing his belief in the positive lifestyle with the 350 employees of Bonnie Bell, Inc. Today the office employees in Lakewood, OH, a Cleveland suburb, have an exercise room with a shower and locker facility and use a park adjacent to the office for jogging, running, biking, and games. At the manufacturing plant in nearby Westlake there are tennis courts and a running track, as well as exercise rooms and showers, and at both units the employees' dining room offers a salad bar, and the half-hour lunch break has been upgraded to an hour. "That way, our people can exercise for 45 minutes, and then have a healthy meal," Bell said. In fact, office employees often wear their warm-up clothes at their desks. "At times, the place looks like a gym," he said. That style wouldn't work well in a bank or a boutique, perhaps, but it suits Bell, who has integrated the fitness theme into the company's product line and sponsors a series of Bonnie Bell Women's Races that is featured in Bonnie Bell advertising.

Businessmen who might be inclined to study their fingernails when Bell talks about sponsoring long-distance runs for women snap to attention when he reports what has happened to his business. At a time when health care and health insurance costs have been going through the roof, the cost of health insurance for employees of Bonnie Bell, Inc., has been coming down: For the past five years, the cost has either held steady or declined, Bell said. "Our program has been very cost-efficient in many ways," he added, "including a reduction in absenteeism, increases in productivity and morale, and an improvement in the general well-being of our people."

Track suits at the office aren't everybody's model of business decorum, but lower health insurance costs and higher productivity are something else. These are things you can put in the bank, and bankable assets make businessmen smile.

References

Chapter 1

1 Conrad, C. C. "Employee Fitness: A Worldwide Movement," *Athletic Purchasing and Facilities*, July 1980.

2 Illich, Ivan. *Medical Nemesis: The Expropriation of Health*. New York: Pantheon Books, 1975.

3 McNerney, Walter J. Address presented at the Conference on Future Directions in Health Care: The Dimensions of Medicine, sponsored by the Blue Cross Association, the Rockefeller Foundation, and the Health Policy Program, University of California School of Medicine, San Francisco. New York City, Dec. 10-11, 1975.

4 Breslow, L. and Belloc, N. B. "The Relation of Physical Health Status and Health Practices," *Preventive Medicine*, August 1972.

Chapter 2

1 Zink, Victor. Address presented at a conference on Economics of Health Care sponsored by the Nevada Chapter, Hospital Financial Management Association, Las Vegas, NV, May 1978.

2 Pfeiffer, G., telephone interview, June 1981.

3 DiBona, R. B.; Mortimer, J. D.; and Flanagan, C. L. Background paper for a conference titled A Corporate Approach to Delivery of Health Services to Employees and Their Families, sponsored by the Boston University Program on Public Policy for Quality Health Care, June 3-4, 1977.

4 Pravosudov, V. "The Effect of Physical Exercise on Health and Economic Efficiency," reported in Collis, Martin L., *Employee Fitness*, Ottawa, Canada: Minister of Supply and Services, 1977.

5 Collis, Martin L., op. cit.

6 National Aeronautics and Space Administration — United States Public Health Service, Health Evaluation and Enhancement Program. *American Journal of Cardiology*, November 1972.

Chapter 3

1 Collis, M. L., op. cit.

Chapter 4

1 *Obesity and Health: A Source Book of Current Information for Health Professionals.* U.S. Department of Health, Education, and Welfare, Public Health Service, Division of Chronic Diseases, Heart Disease Control Program, Washington, DC, U.S. Government Printing Office, 1966.

2 Winikoff, B. "Diet Change and Public Policy," in *Future Directions in Health Care: A New Public Policy.* Reports from a conference sponsored by the Blue Cross Association, the Health Policy Program of the University of California Medical School at San Francisco, the Institute of Medicine, and the Rockefeller Foundation, New York, February 1977. Cambridge, MA: Ballinger Publishing Co., 1978.

3 Stamler, J. "Epidemiology and Treatment of Hypertension," in *Future Directions in Health Care: A New Public Policy,* op. cit.

4 Finnerty, F. A. "Use of Paramedical Personnel in Treating Hypertension," in *Future Directions in Health Care: A New Public Policy,* op. cit.

5 Report to the Congress by the Comptroller General of the United States, *Progress and Problems in Treating Alcohol Abusers.* National Institute on Alcohol Abuse and Alcoholism, Department of Health, Education, and Welfare, April 1977.

6 Ibid.

7 Selye, Hans. *The Stress of Life.* New York: McGraw-Hill Book Co., 1956.

8 Selye, Hans. *Stress Without Distress.* New York: New American Library (Signet Books), 1974.

9 Klerman, G. L. "Mental Health: Recommendations for Public Policy," in *Future Directions in Health Care: A New Public Policy,* op. cit.

Chapter 5

1 These and the following observations by Heule and other Control Data Corporation executives were made in interviews with the author at corporate headquarters in Minneapolis, Mar. 19, 1981.

Chapter 6

1 Interview at Skokie Valley Community Hospital, Skokie, IL, Apr. 16, 1981.

2 Mabley, J. "From Brownies to Carrot Sticks: A Community Health Education Study," *Trustee,* August 1979.

3 Interview at headquarters office of the National Assembly of

Baha'is, Evanston, IL, Apr. 16, 1981.

4 *Good Health Program, Statement of Objectives*, Skokie Valley Community Hospital, 1979.

Chapter 7

1 Behrens, R. "Climate Ripe for Marketing Strategies," *Hospitals*, Oct. 1, 1979.

2 American Hospital Association, *The Hospital's Responsibility for Health Promotion*, Policy and Statement, 1979.

3 Center for Health Promotion, American Hospital Association, *Hospital-Business Programs*, 1980.

4 *Birth of the Well Being*, Scripps Memorial Hospital Annual Report, La Jolla, CA, 1980.

5 Telephone communications, May 1981.

6 MacKenzie, J. "Keeping Workers Healthy," *San Francisco Business*, May 1979. Reprinted by the Health Promotion Center, American Hospital Association.

7 Telephone communication, May 1981.

8 Letter to the author, May 12, 1981.

9 Allen, R. F. with Linde, S. *Lifegain: The Exciting New Program That Will Change Your Health — and Your Life*. New York: Appleton-Century-Crofts, 1981.

10 Allen, R. F.; Adamson, G.; Bellingham, R.; and Kraft, C. *Getting Into the Wellness Business: Pitfalls and Potentials for Hospitals*. Morristown, NJ: Human Resources Institute, 1981.

Chapter 8

1 Telephone communication, May 1981.

2 Plan reports to the Blue Cross and Blue Shield Associations.

3 Carlson, D. R. "The Health Engineer: A New Concept for Business and Industry," Proposal for Blue Shield of California, 1981.

4 Blue Cross and Blue Shield Associations, in cooperation with YMCA. *Put Some Pep in Your Step! Walking for Exercise and Pleasure*, Chicago, 1981.

5 Wilson Learning Corporation, *Wilson Wellness*, 1980.

6 Williamson, J. "How to Get Wellness Started," Information document, Wilson Learning Corporation, 1981.

Chapter 9

1 Turner, C. E. *Personal and Community Health*. St. Louis: C. V.

Mosby Co., 1925. Sixth Edition, 1942.

2 Farquhar, J. W. "Community Programs in Control of Hypertension," in *Future Directions in Health Care: A New Public Policy*, op. cit.

3 Maccoby, N. "Strategies for Marketing Health Education," in *Health Education and Promotion: Agenda for the Eighties*, a program for health insurance executives, sponsored by the Health Insurance Association of America and the American Council of Life Insurance, Atlanta, GA, Mar. 16-18, 1980.

4 Wear, R. F. "Existing Health Promotion Programs in Industry," in *Health Education and Promotion: Agenda for the Eighties*, op. cit.

5 Alderman, M. H. "Detection and Treatment of High Blood Pressure at the Work Place," presented at the Secretary's Conference on High Blood Pressure Control in the Work Setting, Washington, DC, Oct. 14, 1976.

6 See Chapter 4, note 4.

7 Green, L. W. *Determining the Impact and Effectiveness of Health Education as It Relates to Federal Policy*, Health Education Monograph 6 (Supplement 1), 1978.

8 Dedmon, R. E. "Kimberly-Clark's Health Management Program: Results and Prospects," Unpublished paper, October 1980.

9 Peepre, Mall. "The Canadian Employee Fitness and Lifestyle Project," in *Health and Fitness: The Corporate View*, Special Report on Employee Fitness in the 1980s, *Athletic Purchasing and Facilities*, 1981.

10 Peters, R.; Benson, H.; and Peters, J. "Daily Relaxation Response Breaks in a Working Population," *American Journal of Public Health*, October 1977.

11 Benson, H. "Relaxation Techniques and Therapies: An Examination of Their Effects on Physiology," Presented at the Conference on Future Directions in Health Care: The Dimensions of Medicine, op. cit.

12 Allen, R. F. Paper Presented to the Society of Prospective Medicine, Stevens Point, Wisconsin, September 1981.

13 Collis, M. L., op. cit.

14 Allen, R. F. The Corporate Buying Spree: Boon or Boondoggle? *S.A.M. Advanced Management Journal*, 45:2 Spring, 1980.

15 Bell, J., telephone interview, June 1981.

Fitness Resources

The following organizations and agencies offer fitness programs such as health assessment services (i.e., the Health Hazard Appraisal and similar systems) for employed groups; most of the listed organizations provide follow-up health education, health promotion, and other fitness programs. This list includes only those organizations that furnished information for this report or were referred to by persons interviewed for this report, and it should be emphasized that listing here does not constitute a recommendation or endorsement of the services offered. Information about these and similar resources can be obtained from the Bureau of Health Education, Center for Disease Control, U.S. Department of Health and Human Services, 1600 Clifton Road, N.E., Atlanta, GA 30333. Information about employee health education and fitness services offered by Blue Cross and Blue Shield Plans may be obtained by contacting your local Plan or the Communications Division, Blue Cross and Blue Shield Associations, 676 North St. Clair Street, Chicago, IL 60611. Finally, information about hospitals offering health assessment and promotion services for employed groups can be obtained from the Center for Health Promotion, American Hospital Association, 840 North Lake Shore Drive, Chicago, IL 60611.

American Health Foundation
320 East 43rd Street
New York, NY 10017

Blue Cross and Blue Shield Associations
676 North St. Clair Street
Chicago, IL 60611

Commonhealth Club
542 College Avenue
Santa Rosa, CA 95404

Control Data Corporation Healthcare Services
8100 34th Avenue South
Minneapolis, MN 55440

General Health
1046 Potomac Street, N.W.
Washington, DC 20007

Health Evaluation
and Longevity Planning Foundation
7300 Fourth Street
Scottsdale, AZ 85251

Health High
930 S. Adams Way
Denver, CO 80209

Human Resources Institute
Tempe Wick Road
Morristown, NJ 07960

Interhealth
Life Extension Institute
2970 Fifth Avenue
San Diego, CA 92103

The Life Management Group
San Diego Sports Medicine Clinic
5555 Reservoir Drive, Suite 100
San Diego, CA 92120

Medical Datamation
Southwest and Harrison Streets
Bellevue, OH 44811

Methodist Hospital of Indiana
Department of Prospective Medicine
1604 N. Capitol Avenue
Indianapolis, IN 46202

Sun Valley Health Institute
P.O.Box 1720
Sun Valley, ID 83353

University of California at San Diego
Extension Division, X-001
La Jolla, CA 92093

University of California at San Francisco
Department of Epidemiology
and International Health
San Francisco, CA 94143

University of Wisconsin Institute
for Lifestyle Improvement
Stevens Point, WI 54481

Wellness Centers of America, Inc.
13 Office Park Circle
Birmingham, AL 35223

Wellness Resource Center
42 Miller Avenue
Mill Valley, CA 94941

Wilson Learning Corporation
6950 Washington Avenue South
Eden Prairie, MN 55344

Wisconsin Center for Health Risk Research
University of Wisconsin
600 Highland Avenue
Madison, WI 53792

Index

About The Author

Robert M. Cunningham, Jr., is a consultant to the Blue Cross and Blue Shield Associations and a contributing editor of *Hospitals* and *Trustee* magazines. He was formerly editor of *Modern Hospital* and *Modern Nursing Home* magazines and publisher of McGraw-Hill's institutional publications in Chicago. He is a graduate of the University of Chicago and an Honorary Fellow of the American College of Hospital Administrators, and he has written several books on hospital and medical affairs.